ALLE ZEIT WACH
1842

44. Colloquium der Gesellschaft für Biologische Chemie
22.-24. April 1993 in Mosbach/Baden

Glyco-and Cellbiology

Biosynthesis, Transport, and Function of Glycoconjugates

Edited by F. Wieland and W. Reutter

With 66 Figures

Springer-Verlag

Berlin Heidelberg New York
London Paris Tokyo
Hong Kong Barcelona
Budapest

Professor Dr. Felix Wieland
Universität Heidelberg
Institut für Biochemie I
der Universität
Im Neuenheimer Feld 328
D-69120 Heidelberg, FRG

Professor Dr. med. Werner Reutter
Freie Universität Berlin
Institut für Molekularbiologie
und Biochemie
Arnimallee 22
D-14195 Berlin, FRG

ISBN-13: 978-3-642-78731-7 e-ISBN-13: 978-3-642-78729-4
DOI: 10.1007/978-3-642-78729-4

Library of Congress Cataloging-in-Publication Data. Gesellschaft für Biologische Chemie. Colloquium (44th: 1993: Mosbach, Baden-Württemberg, Germany) Glyco- and cellbiology: biosynthesis, transport, and function of glycoconjugates: 44. Colloquium der Gesellschaft für Biologische Chemie 2.-24. April 1993 in Mosbach/Baden / edited by F. Wieland and W. Reutter. p. cm. Includes bibliographical references. ISBN-13: 978-3-642-78731-7 (Berlin). -- (New York) 1.Glycoconjugates- Congresses. I. Wieland, F. II. Reutter, W. III. Title. QP702.G577G47 1993 574.19'2483-dc20 93-47463

Typesetting: M. Masson-Scheurer, Homburg/Saar
39/3130-5 4 3 2 1 - Printed on acid-free paper

Contents

Contributors

You will find the addresses at the beginning of the respective contribution

Components and Mechanisms Involved in Protein Translocation Through the ER Membrane

T. A. Rapoport[1], D. Görlich[1], E. Hartmann[1], S. Prehn[2], and K.-U. Kalies[1]

1 Introduction

Many proteins are transported through the ER membrane as they are synthesized. These include secretory proteins and proteins of the plasma membrane, of lysosomes, and of all organelles of the secretory pathway. Synthesis of these proteins begins in the cytoplasm, but they are then targeted to the ER membrane by signal sequences, which are characterized by a stretch of at least six apolar amino acids, and which are often located at the amino terminus of precursor molecules. Recognition of the signal sequence and targeting of the nascent chain generally requires the function of the signal recognition particle (SRP) and of its membrane receptor (also called docking protein), but alternative targeting pathways exist. The targeting phase is followed by the actual translocation process. Proposed mechanisms of translocation have ranged from the idea that the transport of a polypeptide chain occurs directly through the phopholipid bilayer without participation of membrane proteins, to models according to which polypeptides are transported through a hydrophilic or amphiphilic channel formed from transmembrane proteins (for discussion, see Rapoport 1991). It now appears that a protein-conducting channel does exist. The evidence comes from electrophysiological data and from the identification of membrane proteins as putative channel constituents. The present paper summarizes our knowledge on the components involved in the actual translocation process. The mechanisms by which polypeptides are targeted to the ER membrane have been discussed elsewhere (Rapoport 1992).

Polypeptides are transported at specific sites through the ER membrane (translocons) that are probably rather complex structures, consisting of a number of proteins with different functions. Some of the components of the translocation site may be directly involved in the transport process, others may take part in chemical modifications of a nascent polypeptide or in its folding and assembly. The complexity of the translocation site is indicated by the number of components that have been found in it (Table 1).

Evidence for a protein-conducting channel has been provided using electrophysiological methods (Simon and Blobel 1991). Channels of high ion conductivity were observed after fusion of rough microsomal vesicles into planar lipids. The channels increased in number after release of the nascent chains from the ribosomes by puromycin, suggesting that they had been plugged by nascent chains in transit through the

[1] Max-Delbrück-Center for Molecular Medicine, Robert-Rössle-Straße 10, D-13122 Berlin-Buch, FRG.
[2] Institute of Biochemistry, Humboldt-University, Hessische Straße 3–4, D-10115 Berlin, FRG.

44. Colloquium Mosbach 1993
Glyco- and Cellbiology

Table 1. Components of the translocon

Protein	Function
Sec61/SecYp	Constituent of a protein conducting channel
TRAM protein	Early function in translocation
Sec62/63p-complex	Early function in translocation in yeast
TRAP-complex	?
Signal peptidase complex	Signal peptide cleavage
Oligosaccharyl transferase	Asn-glycosylation

ER membrane. They closed if the salt concentration was subsequently increased – conditions known to result in the dissociation of the ribosomes into their subunits. Further evidence for a hydrophilic channel comes from experiments in which the environment of membrane-inserted nascent chains was investigated by measuring the fluorescence life-time of incorporated fluorescent probes (A. E. Johnson, pers. comm.).

2 Sec61/SecYp: the Major Component of the Protein-Conducting Channel

Sec61 was discovered in *S. cerevisiae* in genetic screens for translocation defects (Deshaies and Schekman 1987). Temperature-sensitive mutations in Sec61p led to the accumulation of precursor molecules of exported proteins at nonpermissive temperatures.

A mammalian homologue of Sec61p was discovered recently (Görlich et al. 1992a); 56% of the amino acids are identical with those of the yeast protein. Sec61p can be crosslinked to various translocating secretory proteins in ER membranes from *S. cerevisiae* (Sanders et al. 1992; Müsch et al. 1992) or canine pancreas (Görlich et al. 1992a). At late stages of the translocation process, when the nascent chain has a sizable lumenal domain, Sec61p is its major crosslinking partner. Thus, Sec61p seems to be closely apposed to polypeptides that are moving through the membrane.

Sec61p has sequence similarity to SecYp, a key component of the protein export machinery of bacteria (Görlich et al. 1992a, Schatz and Beckwith 1990) (Fig. 1). The proteins have identical predicted topologies. Several hydrophilic amino acids within membrane-spanning regions are conserved, suggesting that they are essential for a hydrophilic environment within the membrane.

Sec61p has many of the properties expected for a constituent of a protein-conducting channel: (1) it is a neighbor of translocating nascent chains in different organisms; (2) its structure suggests that it may form a hydrophilic environment in the membrane; (3) it is highly conserved in evolution; and (4) Sec61p from yeast and its prokaryotic counterpart SecYp are essential for translocation in vivo, Sec61p from mammals is required for in vitro translocation in a reconstituted system.

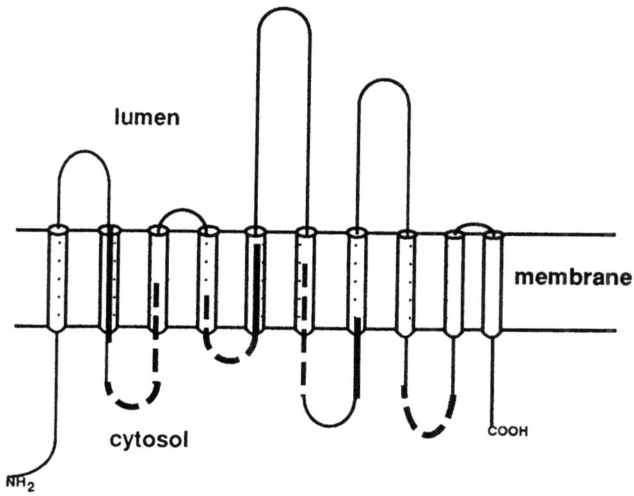

Fig. 1. Predicted membrane topologies of Sec61p and SecYp. *Thick regions* indicate sequences with conserved amino acids and high similarity between Sec61p and SecYp from various bacteria, *dashed regions* indicate sequences of lower but significant similarity

3 Ribosome Association of Sec61p

Mammalian Sec61p is tightly bound to ribosomes after solubilization of rough microsomes with detergent at high salt concentrations (Görlich et al. 1992a). The interaction cannot occur exclusively via the nascent chain because the latter can be released from the ribosome by puromycin without causing detachment of Sec61p. However, if the ribosome is dissociated by increasing the salt concentration, Sec61p is released. The conditions required to strip rough microsomes from ribosomes are the same as those needed for the dissociation of the isolated Sec61p-ribosome complex. These results are consistent with those from the electrophysiological experiments that provide evidence for a protein-conducting channel. It appears that the closure of the channel is caused by the dissociation of the ribosomal subunits from Sec61p during termination of translation. The physiological relevance of the interaction between Sec61p and ribosomes is also supported by the fact that the interaction is induced by the targeting of a nascent polypeptide chain to the ER membrane (Görlich et al. 1992a).

Because Sec61p is probably a core component of the translocation site, it is likely that its association with the ribosome is caused by a ribosome receptor. The latter may be either Sec61p itself or an associated "adaptor" protein. Recent experiments have shown that Sec61p exists as a complex with two smaller polypeptide chains (Görlich et al. unpubl.).

A 180-kDa protein (Savitz and Meyer 1990) and a 34-kDa protein (Tazawa et al. 1991) have been proposed as ribosome receptors before, but counterarguments have been raised against both candidates (Nunnari et al. 1991; Görlich et al. 1992a).

The interaction of membrane-bound ribosomes with Sec61p indicates that nascent chains are transferred directly from the channel in the ribosome into a protein-conducting channel in the membrane. This would prevent the premature folding of a polypeptide chain in the cytoplasm that may prevent subsequent translocation. However, at least in yeast, Sec61p is also involved in the posttranslational translocation of some proteins, such as the secretory protein prepro-α-factor (Sanders et al. 1992; Müsch et al. 1992). Presumably, these proteins maintain a translocation competent state even after their release from the ribosome.

4 The TRAM Protein

Another component of the translocon is the TRAM protein (Görlich et al. 1992b). Its identification was based on the work of Nicchitta and Blobel (1990), who reconstructed translocation activity in proteoliposomes, and studies that showed that short nascent chains of a secretory protein can be crosslinked to a glycosylated membrane protein (Wiedmann et al. 1987; Krieg et al. 1989). For purification of the crosslinking partner, proteoliposomes, with a defined composition of glycoproteins were reconstituted from a detergent extract of canine pancreas microsomes, and tested for the appearence of a crosslinked product. A single glycoprotein, the "*tr*anslocating chain *a*ssociating *m*embrane (TRAM)" protein, was sufficient to allow crosslinking (Görlich et al. 1992b). The sequence of the TRAM protein, deduced from cloning of the corresponding cDNA, suggests that it spans the membrane eight times and that it has a cytoplasmic tail of about 60 amino acids (Fig. 2). Several amino acids in the membrane-spanning regions are hydrophilic or charged.

The effect of the TRAM protein on the translocation of secretory proteins was tested in an improved reconstitution system (Görlich et al. 1992b). Proteoliposomes depleted of glycoproteins had a reduced transport activity for prepro-α-factor and pre-β-lactamase but had an only slightly reduced activity for preprolactin. Replenishment of the proteoliposomes with the TRAM protein was sufficient to restore translocation to the original levels.

The differential effect of glycoprotein depletion on different translocation substrates is similar to the effect of several Sec mutants (Deshaies and Schekman 1989),

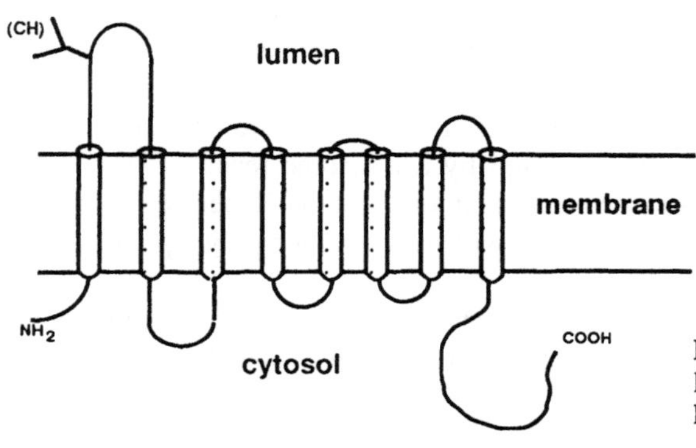

Fig. 2. Predicted membrane topology of the TRAM protein. *CH* Asn-linked carbohydrate chain

and may indicate different requirements for translocation components.

Short nascent chains of secretory proteins representing early stages of translocation can be crosslinked to both the TRAM protein and Sec61p (Görlich et al. 1992a). The TRAM protein seems to interact with amino acid residues preceding the hydrophobic core of the signal sequence whereas Sec61p seems more in contact with amino acids in the hydrophobic core and with residues succeding it (High et al. 1993).

With longer chains of secretory proteins, crosslinks to the TRAM protein have not been observed. It is possible that these longer chains lack suitable located amino acids for crosslinking, but it seems more likely that the TRAM protein is only adjacent to nascent chains at the beginning of their membrane passage. It seems possible that signal peptide cleavage causes displacement of the TRAM protein.

5 The Sec62p-Sec63p Complex

Sec62p and Sec63p were detected similarly to Sec61p in *S. cerevisiae* in genetic screens for translocation components (Deshaies and Schekman 1987). They form a complex that also includes a glycoprotein of 31.5 kDa and a nonglycoprotein of 23 kDa (Deshaies et al. 1991). Sec63p interacts with BiP (Kar2), a chaperone located in the lumen of the ER. Because temperature-sensitive mutations of BiP result in the rapid appearance of translocation defects at nonpermissive temperatures (Vogel et al. 1990), it seems possible that the chaperone directly participates in the transport process of at least some proteins, perhaps by pulling the polypeptide chain across the membrane.

The Sec62p-Sec63p complex may function at early stages of the translocation process during which nascent polypeptides give weak crosslinks to Sec62p (Müsch et al. 1992). Mutations in Sec62p or Sec63p also prevent the interaction of translocating chains with Sec61p (Sanders et al. 1992).

6 Enzymes in the Translocation Site

Two enzymes are known to be located in the translocon: the signal peptidase and the oligosaccharyltransferase. They catalyze cotranslational modifications of the polypeptide chain and are unusual enzymes in that they are as abundant as their substrates. A special structural arrangement is probably needed for the enzymes to act on uncompleted nascent chains which may be surrounded by other membrane proteins.

The signal peptidas consists of five different proteins (Evans et al. 1986). The oligosaccharyltransferase has been purified from dog pancreatic microsomes and consists of three subunits, the two ribophorins (I and II) and a 48-kDa polypeptide (Kelleher et al. 1992).

Both the oligosaccharyltransferase and the signal peptidase seem to be dispensable for the actual translocation process because protein translocation occurs in proteoliposomes depleted of all glycoproteins (including both enzymes) except the TRAM protein (Görlich et al. 1992b).

7 The Translocon-Associated Protein (TRAP) [Previously Called Signal Sequence Receptor (SSR)] Complex

The signal sequence receptor α subunit (SSRα) (recently renamed TRAPα) was purified on the basis of its properties as deduced from its crosslinking to short translocating polypeptide chains (Hartmann et al. 1989). The nascent chains are crosslinked to a protein that is about 35 kDa, has a cytoplasmic tail of about 5 kDa, is glycosylated, and is abundant (Wiedmann et al. 1987). Both the TRAM protein and TRAPα meet these requirements, and the TRAM protein is actually the major crosslinking partner of short nascent chains (Görlich et al. 1992b). However, TRAPα can also be crosslinked to various translocating chains, and the proportion of TRAPα among the glycoproteins crosslinked to the nascent chain appears to increase as chain length increases (Görlich et al. 1992b). The protein is not a signal sequence receptor (therefore the change of the name) and most likely not even directly involved in translocation. Proteoliposomes reconstituted from a detergent extract from which TRAP had been removed by immuno-affinity chromatography (Migliaccio et al. 1992), and proteoliposomes containing the TRAM protein as the only glycoprotein (Görlich et al. 1992b) both have unimpaired translocation activity.

Nevertheless, the TRAP complex is likely to be located in the translocation site. It is associated in part with ribosomes after solubilization of rough microsomes, it is segregated to the rough portion of the ER, and can be crosslinked to membrane-bound ribosomes (for discussion, see Rapoport 1991). Antibodies to TRAPα and Fab frag-

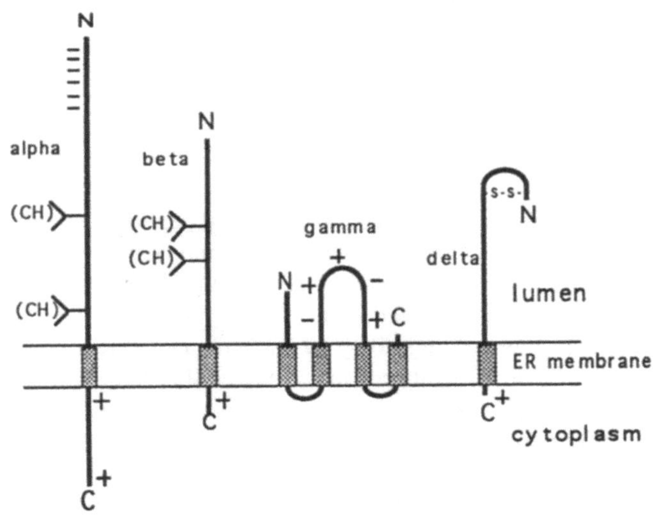

Fig. 3. Predicted membrane topologies of the subunits of the TRAP complex. CH Carbohydrate chains. *Plus* and *minus signs* indicated positively and negatively charged polypeptide segments

ments prepared from the antibodies inhibit the in vitro translocation of several secretory proteins (Hartmann et al. 1989).

TRAPα is a constituent of a stoichiometric complex of four membrane proteins (TRAP complex). The amino acid sequences deduced from cloning of the corresponding cDNAs indicate that the α, β, and δ subunits span the membrane only once. The γ subunit is predicted to span the membrane four times (Fig. 3).

The function of the TRAP complex remains unclear. It may have an enzymatic activity or it could be required for the translocation of only a subclass of proteins. Alternatively, it could function to clear the translocation site or to facilitate folding and assembly of translocated proteins.

8 Conclusions

It now seems certain that membrane proteins are instrumental in the translocation process. The translocon contains not only proteins that are essential for the transport process, but also enzymes that catalyze the modification of nascent polypeptides and probably proteins needed for other functions (see Table 1). A protein-conducting channel is likely to exist.

Sec61/SecYp seems to be the major component of the putative protein-conducting channel. Other membrane proteins, like the TRAM protein or the Sec62p-Sec63p complex may be only transiently involved and perhaps not even needed for all proteins. In general, at least in mammals, the nascent polypeptide seems to be transferred directly from the ribosome into the translocation site; this is made possible by a tight junction between the membrane-bound ribosome and Sec61p.

The mechanisms of protein transport across the ER membrane and across the cytoplasmic membrane in bacteria seem to be basically the same. The similarity between SecYp and Sec61p provides a mechanistic correlate to the long known fact that signal sequences are similarly structured and exchangable between proteins of different classes of organisms.

The precise mechanism of translocation remains unclear but one may now hope to reconstitute into proteoliposomes the translocation process from purified membrane components.

References

Deshaies RJ, Schekman R (1987) A yeast mutant defective at an early stage in import of secretory protein precursors into the endoplasmic reticulum. J Cell Biol 105:633–645

Deshaies RJ, Schekman R (1989) Sec62 encodes a membrane protein required for protein translocation into the yeast endoplasmic reticulum. J Cell Biol 109:2653–2664

Deshaies RJ, Sanders SL, Feldheim DA, Schekman R (1991) Yeast Sec proteins involved in translocation into the endoplasmic reticulum are assembled into a membrane-bound multisubunit complex. Nature 349:806–808

Evans E, Gilmore R, Blobel G (1986) Purification of microsomal signal peptidase as a complex. Proc Natl Acad Sci USA 83:581–585

High S, Martoglio B, Görlich D, Andersen SSL, Ashford AJ, Giner A, Hartmann E, Prehn S, Rapoport TA, Dobberstein B, Brunner J (1993) Site specific photocrosslinking reveals that Sec61p and TRAM contact different regions of a membrane inserted signal sequence. Submitted for publication

Kelleher DJ, Kreibich G, Gilmore R (1992) Oligosaccharyltransferase activity is associated with a protein complex composed of ribophorins I and II and a 48 kd protein. Cell 69:55–65

Krieg UC, Johnson AE, Walter P (1989) Protein translocation across the endoplasmic reticulum membrane: identification by photocrosslinking of a 39-kD integral membrane glycoprotein as part of a putative translocation tunnel. J Cell Biol 109:2033–2043

Migliaccio G, Nicchitta CV, Blobel G (1992) The signal sequence receptor, unlike the signal recognition particle receptor, is not essential for protein translocation. J Cell Biol 117:15–25

Müsch A, Wiedmann M, Rapoport TA (1992) Yeast Sec proteins interact with polypeptides traversing the endoplasmic reticulum membrane. Cell 69:343–352

Nicchitta C, Blobel G (1990) Assembly of translocation competent proteoliposomes from detergent-solubilized rough microsomes. Cell 60:259–269

Nunnari JM, Zimmerman DL, Ogg SC, Walter P (1991) Characterization of the rough endoplasmic reticulum ribosome-binding activity. Nature 352:638–640

Rapoport TA (1991) Protein translocation across the endoplasmic reticulum membrane: facts, models, mysteries. FASEB J. 5:2792–2798

Rapoport TA (1992) Transport of proteins across the endoplasmic reticulum membrane. Science 258:931–936

Sanders SL, Whitfield KM, Vogel JP, Rose MD, Schekman RW (1992) Sec61p and BiP directly facilitate polypeptide translocation into the ER. Cell 69:353–366

Savitz AJ, Meyer DI (1990) Identification of a ribosome receptor in the rough endoplasmic reticulum. Nature 346:540–544

Schatz OJ, Beckwith J (1990) Genetic analysis of protein export in Escherichia coli. Annu Rev Genet 24:215–248

Simon SM, Blobel G (1991) A protein-conducting channel in the endoplasmic reticulum. Cell 65:371–380

Tazawa S, Unuma M, Tondokoro N, Asano Y, Ohsumi T, Ichimura T, Sugano H (1991) Identification of a membrane protein responsible for ribosome binding in rough microsomal membranes. J Biochem 109:89–98

Vogel JP, Misra LM, Rose MD (1990) Loss of BiP/GRP78 function blocks translocation of secretory proteins in yeast. J Cell Biol 110:1885–1895

Wiedmann M, Kurzchalia TV, Hartmann E, Rapoport TA (1987) A signal sequence receptor in the endoplasmic reticulum membrane. Nature 328:830–833

Protein Sorting Signals for Maintaining Transmembrane Proteins in the Endoplasmatic Reticulum

M. R. Jackson and P. A. Peterson[1]

1 Introduction

Trafficking and sorting of proteins through the central organellar system has been a focus of interst since Palade (1975) and co-workers formulated the general outline of the secretory pathway. The basics of this pathway are now well established. Nascent proteins are targeted to the first organelle of this pathway, the endoplasmic reticulum, via a signal sequence (Blobel and Dobberstein 1975). Subsequently, proteins are transported in a vectorial manner between the organelles by a series of vesicle budding and fusion events. Despite the extensive movement of both protein and lipid into and out of them, each organelle is maintained as an apparently discrete membrane-bounded unit. An organelle in this pathway is thus required simultaneously to allow transit of proteins through it, and to maintain a set of resident proteins which presumably define its structural and functional properties. We have been interested to determine how these resident proteins are sorted from those destined for transport to the next compartment. This information is likely to be the key to understanding how the integrity of the organelles is maintained in such a dynamic environment. In particular we have focused on the mechanism by which resident membrane proteins of the endoplasmic reticulum are maintained in this organelle.

The long-standing controversy over whether trafficking and sorting in the secretory pathway rely on either retention or transport signals (see Rothman 1987; Klausner 1989) now seems like a desire to oversimplify the system. Multiple mechanisms of retention of proteins in the ER have been described and although specific signals for transport have not been identifed, it is probably more appropriate to recognize that proteins need to acquire a transport-competent state before they can leave the ER. In recent years, it has become clear that the ER is one of the primary quality control stations, retaining proteins that have not folded correctly or have not reached their mature quaternary structure (see Rose and Doms 1989; Bonafacino et al. 1990). Nevertheless, upon attaining a mature correctly folded state, current opinion favors the idea that transport of proteins between the organelles occurs by default (Wieland et al. 1987; Pfeffer and Rothman 1987). Consequently, residency in a specific organelle of this pathway requires mechanisms that recognize and maintain the proteins in that location (Pelham 1989).

We have identified short discrete signals in the cytoplasmic tails of ER resident membrane proteins which maintain proteins in the ER (Nilsson et al. 1989; Lotteau et

[1] Department of Immunology, IMM8, The Scripps Research Institute, 10666, North Torrey Pines Road, La Jolla, California 92037 USA.

44. Colloquium Mosbach 1993
Glyco- and Cellbiology

al. 1990; Jackson et al. 1990). This chapter describes the characteristics of these signals and our progress in understanding how they may function.

2 ER Targeting Motifs of Type I and III Trans-Membrane Proteins

We have previously shown that the last six amino acids (DEKKMP) of the short cytoplasmic tail of the ER resident type I membrane protein E19, a 19 kDa protein encoded by Adenovirus 3, are both necessary and sufficient for ER targeting (Nilsson et al. 1989). Transplanting the cytoplasmic tail sequence of the E19 protein onto marker proteins, e.g., the T cell surface glycoproteins CD8 or CD4, resulted in efficient targeting of the chimeric proteins (CD8/E19 and CD4/E19) to the ER. Screening of the carboxy-terminal sequences of other known ER resident type I and III membrane proteins for their ability to maintain CD8 in the ER (Jackson et al. 1990) allowed us to identify a lysine-positioned three residues from the carboxy-terminus (– 3) as a common feature of the sequences that resulted in ER targeting (see Table 1). Mutation of this lysine to a serine (– 3S mutants) in each of these sequences destroys the targeting motif, suggesting that all these sequences operate by the same mechanism. Further mutation analyses of CD8/E19 implied a second lysine in the – 4 position as the only other essential residue of this ER targeting motif. The – 4 lysine could, however, be moved to the – 5 position without destroying the signal. The critical positions of the lysines were also demonstrated by analyses of addition and deletion mutants. While all such changes resulted in cell surface expression of the marker protein, the rate of transport was found to be directly proportional to the extent of the deletion from, or addition to, the E19 sequence (Jackson et al. 1990).

The finding that two lysines in defined positions represent the critical elements of this motif was confirmed by our ability to create an ER targeting motif in a novel sequence context. Thus, introduction of a lysine residue adjacent to the – 3 lysine in the T cell surface protein Lyt 2 rendered this molecule, Lyt2-4K, a resident of the ER. This result prompted us to characterize this motif in the context of a defined sequence. Using a poly-serine tail as the neutral background, lysines were introduced in various positions. The incorporation of a lysine diminished the transport rate by various extents, depending on the position; as expected, lysine in the – 3 position had the most profound effect, followed by lysines in the – 4 and – 5 positions. The simultaneous introduction of two lysine residues considerably reduced the transport rate of the marker protein, but only when the lysines occupied positions – 3 and – 4, or – 3 and – 5 were molecules prevented from reaching the cell surface.

As the list of sequences of resident ER membrane proteins grows, the generality of this motif is becoming apparent. While the basic consensus we described still holds (see Table 1) some substitutions of lysine by arginine are obviously permitted. However, throughout this chapter, we refer to this motif as the double lysine or KK motif.

To date, more than ten different families of ER membrane proteins, both type I and type III, have now been identified with this motif, from species as diverse as yeast and man. Some proteins, e.g., SSRβ, gp25L, and TRAM are recognized as residents of the rough ER, while others, e.g., HMG CoA reductase and UDPGT, are

Table 1. Carboxy-terminal sequences of ER resident membrane proteins

Ad2 E19	F I D E **K K** M P	(1)	SSRβ (gp25H)	D T P **K** S **K K** N	(12)
Ad5 E1	F I E E **K K** M P	(2)	gp25L	F F I A **K K** L V	(13)
Ad3 E19	N E E **K** E **K** M P	(3)	Calnexin	R N R **K** P **R R** E	(13)
UDPGT H25	T G K **K** G **K** R D	(4)	TRAM	R N R **K** E **K** S S	(14)
UDPGT H4	K G K **K K K** R D	(5)	I.C p53	A A A A **K K** F F	(15)
UDPGT HP1	K A H **K** S **K** T H	(6)	OST48	M K E **K** E **K** S D	(16)
UDPGT R23	K E K **K** M **K** N E	(7)	Yeast p45	L E T F **K K** T N	(17)
UDPGT 2F	N M G **K K K** K E	(8)			
UDPGT K39	K G H **K** S **K** T H	(9)	Consensus		
HMG CoA	G T C T **K K** S A	(10)	3/4K	- - - - K K - -	
SER 53KD	P K N **R** Y **K** K H	(11)	3/5K	- - - K - K - -	

References: (1) Hérissé et al. 1980 (2) Signäs et al. 1986 (3) Cladaras and Wold 1985 (4) Jackson et al. 1987 (5) Jackson and Burchell 1986 (6) Harding et al. 1988 (7) Jackson and Burchell 1986 (8) Jackson and Burchell 1986 (9) Iyanagi et al. 1986 (10) Luskey and Stevens 1985 (11) Leberer et al. 1989 (12) Gorlich et al. 1990 (13) Wada et al. 1991 (14) Gorlich et al. 1992 (15) Itin et al. 1993 (16) Silberstein et al. 1992 (17) te Heesen et al. 1991.

primarily localized to the smooth ER. Furthermore, p53 a marker protein of the intermediate compartment, which has been suggested to recirculate between the ER and the cis-Golgi, also possesses a double lysine motif (Hauri and Schweizer 1992).

3 ER Targeting of Type II Membrane Proteins

The invariant chain (Ii) protein is a type II membrane protein. Use of alternative translation initiation sites in human invariant chain results in the synthesis of two polypeptides, Iip33 and Iip31 (Strubin and Long 1986). Iip33, is an ER resident protein with a cytoplasmic tail of 45 residues, while Iip31, which has a cytoplasmic tail of 30 residues, is transported out of the ER and directs the intracellular targeting of MHC class II to endosomes (Lotteau et al. 1990). These data suggest that the 15 amino terminal residues on Iip33 encode an ER targeting motif.

Using a cassette mutagenesis system similar to the one we devised for characterizing the double lysine motif, we have generated many different substitution mutations in the p33 amino terminal sequence (Schutze et al. 1994.).

Based on such mutants and others, it is clear that the ER targeting motif of Iip33 is composed of arginine residues located close to the amino terminus, in much the same way as the ER targeting motif of type I membrane proteins is based on lysine residues close to the COOH terminus. The essential component of the motif is two arginine residues located either adjacent to one another at positions 2 and 3, 3 and 4, or 4 and 5, where position 1 is the initiator methionine. The two arginine residues may also be split by a single residue, i.e., 2 and 4 and 3 and 5. Throughout this chapter, we refer to this motif as the double arginine motif or, for convenience, RR. As for the KK motif, a terminal postion is essential for this RR motif. If it is placed further away than position 4 and 5, it no longer functions. In some postions, but only in specific combinations, R may be replaced by K without affecting ER targeting. For example, 3R/4K and 4R/5K are functional motif, whereas 3K/5R does not function. These characteristics are again similar to the KK motif, where it is now recognized that R may in some instances replace K. However, we have found no case where a motif composed only of lysines will function in place of an RR motif to target type II membrane proteins; similarly, a motif composed only of arginines cannot functionally replace a KK motif to target type I membrane proteins. The motifs are thus extremely similar in their general requirements, but at the same time distinct in the precise combination of which basic residues are functional.

In order to demonstrate that the RR motif is sufficient for targeting of a protein to the ER, the amino terminal 5 residues of human transferring receptor (TfR), a type II membrane protein which is ordinarily expressed at the cell surface, were replaced by those from Iip33. The resulting recombinant TfR-RR, when expressed in HeLa cells, localized exclusively to the ER by immunofluorescence microscopy. A search of the protein data base for membrane proteins whose cytoplasmic tail sequences fit the RR consensus identified in addition to Iip33, TRAM, an ER resident protein involved with protein translocation (Gorlich et al. 1992). This protein has eight trans-membrane spanning domains and both its amino-and carboxy-termini are located in the cytoplasm. The amino terminal sequence is NH2 –MAI**RKK**STKS (we show above that 4R/5K is a functional motif on Ii), while the COOH sequence is RN**RKEK**SS-COOH, raising the possibility that this protein may be maintained in the ER by both KK and RR motifs.

4 How Do the KK and RR Motifs Maintain Proteins in the ER?

In principle, two mechanisms can be envisaged as to how the KK and RR motifs function. Either these motifs retain (fix) proteins in the ER or they function by retrieving proteins from post-ER compartments in a manner analogous to that suggested for ER resident soluble proteins with the KDEL motif (Pelham 1989). We have carried out extensive analyses of the subcellular localization and posttranslational modifications of marker protein maintained in the ER by KK, RR, and KDEL motifs, looking for evidence that might indicate which of these two mechanisms is utilized (Jackson et al. 1993). We have concentrated our efforts on understanding targeting by the KK motif, in the belief that the RR motif most likely functions in an analogous fashion.

4.1 KK and RR Tagged Proteins Access Post-ER Compartments

Marker proteins tagged with the double lysine motif, e.g., CD8/E19 and CD4/E19, were found to rapidly receive posttranslational modifications characteristics of the intermediate compartment (Bonnatti et al. 1989; Tooze et al. 1989), e.g., GalNAc and palmitate, and partially localized to this organelle by immunofluorescence microscopy. Similarly, TfR-RR was found to partially localize to the intermediate compartment. Furthermore, KK tagged chimeras were also found to receive characteristic

Golgi modifications, i.e., CD8/E19 received sialic acid and galactose, whereas the N-linked sugars on CD4/E19 became partially endo H-resistant. However, these Golgi modifications occurred very slowly over a 24-h period and in a stepwise fashion.

4.2 Golgi Modified KK and KDEL Tagged Proteins Localize to the ER

The above data indicate that proteins accumulated in the ER as a result of an RR or KK motif have access to post-ER compartments. However, they do not prove that these escaped proteins are subsequently retrieved to the ER. To demonstrate this, we have analyzed whether the ER of cells expressing high levels of KK CD8 chimeras stain with lectins (Piller et al. 1989) which recognize specific terminal glycans. *Helix pomatia* (HPA) and *Arachis hypogeae* (PNA) lectins were used to stain for terminal GalNAc and terminal Gal(1–3)GalNAc, respectively (Hammarstrom and Kabat 1969). In cells expressing high levels of CD8/E19, the ER was found to stain strongly with HPA and PNA. A similar result was found for cells expressing high levels of CD8/KDEL, whereas in untransfected cells the lectins primarily stained the Golgi, the expected location for proteins with only partially processed carbohydrate. These lectin-binding data provide the first direct evidence that proteins maintained in the ER by a KK motif are retrieved to the ER from post-ER compartment(s).

4.3 The Context of the KK Motif Affects Exposure to Post-ER Compartments

The ER targeting motif on CD8/E19 or CD4/E19 can be substituted by sequences derived from other ER resident proteins (Jackson et al. 1990). The identification of a KK motif in each of these sequences led us to believe that all these sequences were functionally analogous. However, when the posttranslational modifications and immunolocalization of the various chimeras were analyzed, we were surprised to find that they behaved differently (see Table 2). Whereas CE8/E19 and CD8/HMG CoA partially localized to vesicles that also stained for the intermediate compartment marker protein p58, the other constructs CD8/H25, CD8/HP1, and CD8/53kSER did not localize to these vesicles. A similar phenomenon was observed if these exact same

Table 2. Posttranslational modifications and immunolocalization of various chimeras

Carboxy-terminal sequence		Intermed. Comp staining		Golgi additions		GalNAc[a] (t > 95%)	HPA stain ER	PNA stain ER
		CD8	CD4	CD8	CD4	CD8	CD8	CD8
Ad E19	KYKSRRSKIDE**KK**MP	X	XXX	X	X	~ 4 h	XX	XX
HMG CoA	----- LQGTCT**KK**SA	X	XXX	X	X	~ 4 h	XX	XX
UDPGT H25	----- YRTG**KK**G**KR**D	–	X	–	–	> 8 h	XXX	–
UDPGT H1	----- V**KKAHKSKTH**	–	X	–	–	> 8 h	XXX	–
SER 53K	----- GTP**K**N**R**Y**KKH**	–	X	–	–	> 8 h	XXX	–

[a] Values are chase time required for > 95% of the molecules to obtain a full complement of GalNAc.

cytoplasmic tails were transferred onto CD4. CD4/E19 and CD4/HMGCoA showed much stronger costaining with p58, than CD4/HP1, CD4/H25, and CD4/53kSER. Analysis of the posttranslational modifications of these various chimeras also showed significant differences (Table 2). However, if cells were treated with brefeldin A, all the chimeras received these modifications at about the same rate, suggesting that the differences in the rates at which the different CD8 chimeras acquired O-linked carbohydrate in untreated cells is most likely due to their differential exposure to post-ER transferases rather than differences in enzyme/substrate affinity. Inspection of the sequence of these tails (Table 2), suggests that the overall effectiveness of each motif might be dependent upon the number and position of the lysine (and to a certain extent also arginine) residues, i.e., the more basic the charge, the better, and also on whether the marker protein is a monomer (CD4) or a dimer (CD8).

5 The Double Lysine Motif Directs ER Retrieval

As discussed above, we have shown that CD8/E19 receives GalNAc and palmitate, which are modifications reported (Tooze et al. 1988; Bonnatti et al. 1989) to occur in post-ER location(s). However, these reports could not rule out low levels of ER-located enzymes. Indeed, herein lies one of the fundamental problems with using posttranslational modifications to map the intracellular trafficking of proteins. This problem is particularly acute in the secretory pathway, as the various modifying enzymes and the protein in question are first synthesized into the ER. We have controlled for this problem by showing that a series of CD8 chimeras, all maintaind in the ER, but by double lysine motifs from different ER resident proteins receive post-ER modifications at different rates. We believe that these rate differences reflect differences in the exposure of these proteins to the transferase rather than differences in their ability to act as substrate for ER-located enzymes for the following reasons: the different CD8 chimeras differ only in the sequence of the last ten amino acids of their 15-residue-long cytoplasmic tail. The sequence of their lumenal and transmembrane domains are identical. It would be surprising if these differences were transmitted through the *trans*-membrane portion to alter the folding of the lumenal domain of CD8, the portion the transferases act upon. Three different conformational-specific anti-CD8 antibodies did not detect any differences in either the rate at which the various chimera folded or their final folded state. Further, the various CD8 chimeras received GalNAc and Golgi modifications at the same rate in cells treated with brefeldin A, indicating that they are equally good substrates for the transferases. Perhaps the most convicing argument that the sequences of cytoplasmic tail affect the retrieval process, rather than a conformation of the marker protein, is that transferring the same set of cytoplasmic tails from CD8 to a different marker protein, CD4, also transfers the rank order in both the rates at which the carbohydrates on this marker protein are modified by post-ER enzymes and the degree to which each chimera colocalizes with the p58 intermediate compartment marker protein (Table 2).

Thus, the most plausible explanation for the types and rates of posttranslational modifications of the various ER maintained CD4 and CD8 chimeras is that these modifications are received primarily outside the ER. As the lectin immunofluorescence data show that the these posttranslationally modified proteins are returned to the ER, we conclude that the double lysine motif is a retrieval signal (Jackson et al. 1993).

6 Golgi to ER Retrograde Transport

The most convincing evidence for a Golgi to ER retrograde transport pathway has come from the work of Pelham and colleagues. Their initial observation that KDEL chimeras receive posttranslational modifications consistent with exposure to post-ER compartments (Pelham 1988; Dean and Pelham 1990) led them to propose a KDEL receptor which retrieved KDEL proteins to the ER from post-ER compartments. They have subsequently identified and characterized this receptor, which at steady state is located primarily in the Golgi, supporting its role in retrieval (see Pelham 1989; Lewis and Pelham 1992). Furthermore, over-expression of KDEL chimera was found to result in the relocation of a human KDEL receptor from the *cis*-Golgi to the ER. The mechanism by which the KDEL receptor, which possesses neither a KK nor an RR motif, is retrieved is unknown.

Lumenal ER proteins and their receptors are not the only proteins thought to cycle between ER and Golgi. For example, two proteins (p53 and p58) normally concentrated in an intermediate compartment between the ER and Golgi can be induced to move into the Golgi and then to the ER by appropriate temperature manipulations (Schweizer et al. 1990; Saraste and Svennsson 1991). Treatment of cells with a microtubule-disrupting drug, nocodazole, results in an apparent accumulation of p53 and p58 in the intermediate compartment/Golgi, suggesting that the retrograde transport was facilitated by microtubules. Further support for a retrograde transport route has come from the finding that brefeldin A (BFA), a fatty acid derivative made by several fungi, induces the redistribution of Golgi proteins into the ER via long, tubulovesicular processes extending out of the Golgi along microtubules. Nocodazole, energy poisons, and reduced temperature inhibited this retrograde pathway (Lippincott-Schwartz et al. 1990). In BFA-treated cells, Golgi proteins appeared to cycle between the ER and intermediate compartment marked by the p53 kDa protein. Addition of nocodazole after BFA treatment preferentially inhibited the retrograde movement, causing Golgi proteins to accumulate in the intermediate compartment. Although some reservations remain about the physiological significance of the effects seen with BFA, the body of evidence is now overwhelming that a retrograde pathway exists (Hauri and Schweizer 1992). Indeed, given the massive movement of membrane and proteins out of the ER, some such pathway must exist on theoretical grounds to maintain the total membrane content and surface area of the various organelles in this pathway (Wieland 1987).

7 The Efficiency of Retrieval Depends on the Sequence Context of the KK

Perhaps the most interesting finding from our analysis of the retrieval process was that the rate at which ER-maintained CD8 and CD4 chimeras receive Golgi modifications and whether they localize to the intermediate compartment was found to be dependent upon the sequence context of the double lysine motif and the marker protein used. This suggested that the strength of the retrieval signal is dictating the steady-state location of the chimera. Thus, chimeras which are efficiently retrieved (e.g., where the KK motif is very favorably presented or where there are multiple signals per molecule, note CD8 is a dimer) are localized at steady state almost exclusively to the ER and receive Golgi modifications very slowly, while chimeras which are less efficiently retrieved (e.g., where the KK motif is less favorably presented or the reporter molecule is a monomer, note CD4 is a monomer) localize at steady state to the intermediate compartment/*cis*-Golgi and are more rapidly modified by Golgi enzymes. The steady-state distribution of a protein may thus be envisaged as simply a consequence of the balance of retrograde targeting and anterograde bulk flow.

8 Retrieval May Occur from Multiple Post-ER Compartment

The occurrence of O-linked sialic acid on CD8/E19 and the existence of endo H-resistant forms of CD4/E19 suggest that these proteins have had access to the medial Golgi (Kornfeld and Kornfeld 1985; Farquhar 1985). However, simply altering the sequence context of the KK motif in CD8/E19 or CD4/E19, resulted in proteins that were still accumulated in the ER but which received little or no modifications by Golgi enzymes (Table 2). These data, in conjunction with the different rates of GalNAc addition to the CD8 chimeras (see Table 2), raise the possibility that the various chimeras are retrieved to the ER with different kinetics. Assuming that the forward rates of transport of these proteins are identical, (which is quite possible given that the chimeras differ only in their tail sequences), then one might expect, on average, that retrieval of the different chimeras would occur from different sites in the exocytotic pathway; a view supported by differences in subcellular distribution of the various CD8 and CD4 chimeras (Table 2). We envisage retrieval of proteins to the ER as a continuous process, sorting occurring as soon as the proteins exit the ER and continuing throughout the secretory pathway. By altering the strength (context) or number of retrieval signals, the steady-state distribution of a protein may be altered from almost exclusively ER, e.g., CD8/H25 to almost entirely post-ER, e.g., p53.

Our data are consistent with a model (Fig. 1). whereby the KK and RR motifs are recognized in a stochastic retrieval process using cytosolic components which are not compartmentalized. A similar model was proposed in the distillation hypothesis by Rothman more than 10 years ago (Rothman 1981). In this model he suggests that retrieval of proteins might occur from multiple compartments in the exocytotic pathway to increase the efficiency of the sorting process.

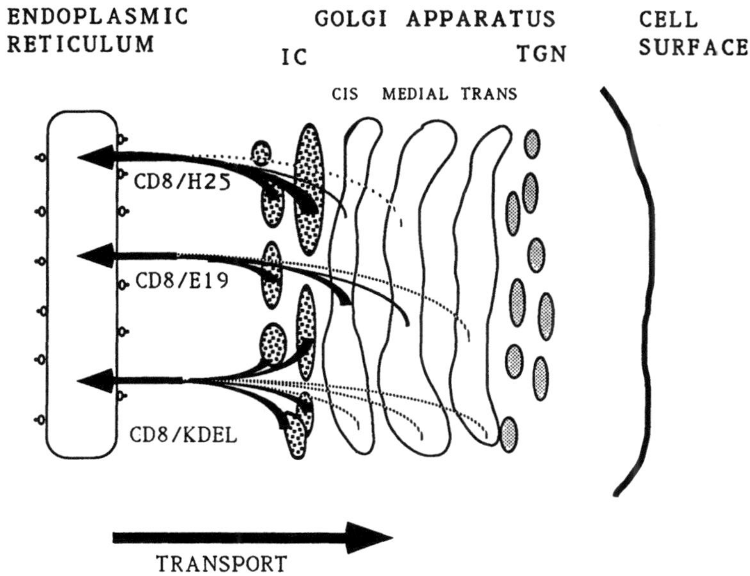

Fig. 1. *Proposed model for the retrieval of CD8/E19, CD8/H25, and CD8/KDEL.* Newly synthesized proteins leave the ER and move to the intermediate compartment (IC), proteins with a strong retrieval motif, e.g., *CD8/H25*, are very efficiently retrieved from this compartment, with minimal leakage to the *cis* Golgi. Proteins with a less efficient retrieval motif, e.g., *CD8/E19*, are retrieved more poorly from the intermediate compartment and are carried along in the anterograde transport to the Golgi stack. At each of the inter-Golgi transport steps protein may be retrieved back to the ER, multiple retrieval operation ensuring minimal leakage of proteins to the cell surface. In this model, KDEL proteins are bound by a KDEL receptor (as suggested by Pelham 1989) most likely in the intermediate compartment and retrieved to the ER. Retrieval of the KDEL receptor is most likely relatively efficient, although some leakage occurs from the intermediate compartment allowing it to patrol and therefore retrieve KDEL proteins from throughout the Golgi

9 Residency in an Organelle by Retention and Retrieval

Although the KK motif is present on many ER resident trans-membrane proteins, several such proteins lack this motif, and it seems likely that mechanisms other than retrieval are responsible for their residency. In contrast to marker proteins tagged with a KK motif, which are slowly modified by Golgi enzymes, endogenous ER proteins lack such modifications (Kornfeld and Kornfeld 1985), as evidenced by the observation that HPA and PNA stain poorly if at all the ER of untransfected cells. Moreover, removal of the KK motif from ER residents such as UDPGT HP1 (M. Jackson unpubl.) does not result in transport of the truncated molecules. These data indicate that the majority of ER-resident transmembrane proteins rarely leave the ER and that they are in some way retained in this organelle by a primary level of ER targeting. Recent analysis of how Golgi enzymes are targeted to this organelle has shown that it is their transmembrane domain that is the essential component for Golgi targeting (see Machamer 1991). The finding that the transmembrane domain of a nuclear pore protein gp210 targets reporter proteins to the pore membrane (Wozniak and Blobel 1993) suggests that targeting to the ER might be achieved, at least in part, by a similar mechanism. Retention in the ER of resident proteins by formation of complexes sufficiently large to be poorly transported has been suggested to explain the observation that proteins of the rough ER form a subdomain resistant to high salt extraction (Horst and Meyer 1985).

In terms of efficiency, it would make much sense if the majority of the ER proteins were prevented from leaving the ER via a retention mechanism. A combination of retention and retrieval would ensure minimal leakage, and that even when overexpressed, a true ER-resident protein is efficiently maintained in the ER. This combination of retention and retrieval could explain why data to substantiate the distillation hypothesis (Rothman 1981) were not previously obtained (Brands et al. 1985). In contrast to true residents, the location of our ER-maintained CD8 and CD4 chimeras is determined only by the apended ER targeting motif. In this respect, these molecules may be unique reagents with which to describe both the biochemical and ultrastructural characteristics of the retrieval pathway.

10 How Do the KK and RR Motifs Direct Sorting to the Retrograde Pathway?

Membrane trafficking is recognized to be directed by highly organized and selective interactions at the cytoplasmic surfaces of membranes. The KK and RR ER targeting motifs thus have prime locations for sorting signals. Selective sorting of these proteins into the retrograde pathway could occur in a manner analogous to the sorting operation utilized in receptor-mediated endocytosis. The KK and RR motifs might then be recognized by adaptin-like molecules (Glickman et al. 1987; Pearse 1988), and thereby concentrated into retrograde vesicles destined for the ER, in much the same way as endocytosis of the LDL receptor is mediated by the sequence motif NPXY, and transferrin receptor by the sequence YXRF (see Chen et al. 1990; Collawn et al. 1990).

It is thus tempting to speculate that a putative family of receptor molecules are recognizing these motifs end on, i.e., the basic residues in the motif and the charged terminus are required for specificity of the interaction. The differences in the sequence requirements of the KK and RR motifs suggest that they are unlikely to be recognized by the same adaptin-like molecule. A di-leucine (LL) endosomal sorting motif has recently been identified in the cytoplasmic tails of the mannose-6-phosphate receptor (Johnson and Kornfeld 1991) and the CD3 γ chain (Letourneur and Klausner 1992). We have also identified a variation of this motif (IL) in the cytoplasmic tail of invariant chain (Schutze and Peterson unpubl.) which is responsible for the targeting of Ii to the endocytic pathway. These findings raise the possibility that a family of adaptin-like molecules that recognize specific pairs of amino acids may exist, these molecules forming the basis for selective sorting of many membrane proteins in exo and endocytic pathways.

Acknowledgments. We are most grateful to Marie-Paul Schutze for helpful comments and for allowing us to cite unpublished data. The work in the author's laboratory was funded by a grants from the National Institute of Health.

References

Blobel G, Dobberstein B (1975) Transfer of proteins across membranes I: presence of proteolytically processed and unprocessed nascent immunoglobulin light chain on membrane-bounded ribosomes of murine myeloma. J Cell Biol 67: 835–851

Bonafacino JS, Cosson P, Klausner RD (1990) Co-localized transmembrane determinants for ER degradation and subunit assembly explain the intracellular fate of TCR chains. Cell 63:503–513

Bonnatti S, Migliaccio G, Simons K (1989) Palmitylation of viral membrane glycoproteins takes place after exit from the ER. J Biol Chem 264:12590–12595

Brands R, Snider MD, Hino Y, Park SS, Gelboin HV, Rothman JE (1985) Retention of membrane proteins by the endoplasmic reticulum. J Cell Biol 101:1724–1732

Chen W-J, Goldstein JL, Brown MS (1990) NPXY, a sequence often found in cytoplasmic tails, is required for ciated pit mediated internalization of the low density lipoprotein receptor. J Biol Chem 264:3116–3123

Cladaras C, Wold WSM (1985) DNA sequence of the early E3 transcription unit of adenovirus. Virology 140:28–43

Collawn JF, Strangel M, Kuhn LA, Esekogwu V, Jing S, Trowbridge IS, Tainer JA (1990) Transferrin receptor internalization sequence YXRF implicates a tight turn in the structural recognition motif for endocytosis. Cell 63:1061–1072

Dean N, Pelham HRB (1990) Recycling of proteins from the Golgi compartment to the ER in yeast. J Cell Biol 111:369–377

Farquhar MG (1985) Progress in unravelling pathways of Golgi traffic. Annu Rev Cell Biol 1:447–448

Glickman JN, Conibear E, Pearse BMF (1989) Specificity of binding of clatherin adaptors to signals on the mannose-6-phosphate/insulin-like growth factor II receptor. EMBO J 8:1041–1047

Gorlich D, Prehn S, Hartmann E, Herz J, Otto A, Kraft R, Wiedmann M, Dobberstein B, Rapoport TA (1990) The signal sequence receptor has a second subunit and is part of translocation complexes in the endoplasmic reticulum as probed by bifunctional reagents. J Cell Biol 111:2283–2294

Gorlich D, Hartmann E, Prehn S, Rapoport TA (1992) A protein of the endoplasmic reticulum involved in early polypeptide translocation. Nature 357:47–52

Hammarstrom S, Kabat EA (1969) Purification and characterization of a blood group reactive hemagglutin from the snail *Helix pomatia* and a study of its combining site. Biochemistry 8:2696–2705

Harding D, Fournel-Gigleux S, Jackson MR, Burchell B (1988) Cloning and substrate specificity of a human phenol UDP-glucuronosyltransferase expressed in COS-7 cells. Proc Natl Acad Sci USA 85:8381–8385

Hauri H-P, Schweizer A (1992) The endoplasmic reticulum-Golgi intermediate compartment. Curr Opin Cell Biol 4:600–608

te Heesen S, Rauhut R, Aebersold R, Abelson J, Aebi M, Clark MW (1991) An essential 45 kDa yeast transmembrane protein reacts with anti-nuclear pore antibodies: purification of the protein, immunolocalization and cloning of the gene. Eur J Cell Biol 56:8–18

Hérissé J, Courtois G, Galibert F (1980) Nucleotide sequence of the EcoRI D fragment of adenovirus 2 genome. Nucl Acids Res 8:2173–2192

Hortsch M, Meyer DI (1985) Immunochemical analysis of rough and smooth microsomes from rat liver: segregation of docking protein in the rough membranes. Eur J Biochem 150:559–564

Itin C, Schindler R, Kappler F, Hauri H-P (1992) Targetting of ERGIC-53 to the ER-Golgi intermediated compartment. J Cell Biochem Abstract supplement 17C, Abstract# H202

Iyanagi TM, Haniu M, Sogawa K, Fuji-Kuriyami Y, Watanabe S, Shiveley JE, Anan KF (1986) Cloning and characterization of cDNA encoding 3-methylcholanthrene inducible rat mRNA for UDP-glucuronosyltransferase. J Biol Chem 261:15607–15614

Jackson MR, Burchell B (1986) The full length sequence of rat liver androsterone UDP-glucuronoslystransferase cDNA and a comparison with other members of this gene family. Nucl Acids Res 14:779–795

Jackson MR, McCarthy LR, Harding D, Wilson S, Coughtrie MWH, Burchell B (1987) Cloning of human liver microsomal UDP-glucuronosyltransferase. Biochem J 242:581–588

Jackson MR, Nilsson T, Peterson PA (1990) Identification of a consensus motif for retention of membrane proteins in the endoplasmic reticulum. EMBO J 9:3153–3162

Jackson MR, Nilsson T, Peterson PA (1993) Retrieval of transmembrane proteins to the endoplasmic reticulum. J Cell Biol 121:317–333

Klausner R (1989) Sorting and traffic in the central vacuolar pathway. Cell 57:703–706

Johnson KF, Kornfeld S (1991) Identification of determinants on the cytolasmic domain of the man-6-P receptors required for efficient lysosomal enzyme sorting. J Cell Biol 115:244a

Kornfeld R, Kornfeld S (1985) Assembly of asparagine-linked oligosaccharaides. Annu Rev Biochem 54:631–664

Lewis MJ, Pelham HRB (1992) Ligand induced redistribution of a human KDEL receptor from the Golgi complex to the endoplasmic reticulum. Cell 68:353–364

Letourner E, Churak JHM, Clarke DM, Green NM, Zubrzycka-Gaarn E, MacLennan DJ (1989) Molecular cloning and expression of cDNA encoding the 53,000 Dalton glycoprotein of rabbit skeletal muscle sarcoplasmic reticulum. J Biol Chem 264:3484–3493

Lippincott-Schwartz J, Donaldson JG, Schweizer A, Berger EG, Hauri H-P, Yuan LC, Klausner RD (1990) Microtubule dependent retrograde transport of proteins into the ER in the presence of brefelin A reveals an ER recycling pathway. Cell 60:821–836

Lotteau V, Teyton L, Peleraux A, Nilsson T, Karlsson L, Schmid SL, Quaranta V, Peterson PA (1990) Intracellular transport of class II MHC molecules directed by invariant chain. Nature 348:600–605

Luskey KL, Stevens B (1985) Human 3-hydroxy-3-methylglutaryl coenzyme A reductase. Conserved domains responsible for catalytic activity and sterol-regulated degradation. J Biol Chem 260:10271–10277

Machamer CE (1991) Golgi retention signals: do membranes hold the key? Trends Cell Biol 1:141–144

Munro S, Pelham HRB (1987) A C-terminal signal prevents secretion of luminal ER proteins. Cell 48:899–907

Nilsson T, Jackson MR, Peterson PA (1989) Short cytoplasmic sequence serve as retention signals for transmembrane proteins in the endoplasmic reticulum. Cell 58:707–718

Palade GE (1975) Intacellular aspects of the process of protein transport. Science 189:347–358

Pearse B (1988) Receptors compete for adaptors found in the plasma membrane coated pits. EMBO J 7:3331–3336

Pelham HRB (1988) Evidence that luminal ER proteins are sorted from secretory proteins in a post-ER compartmen. EMBO J 7:913–918

Pelham HRB (1989) Control of protein exit from the endoplasmic reticulum. Annu Rev Cell Biol 5:1–23

Piller V, Piller F, Klier FG, Fukuda F (1989) O-glycosylation of leukosialin in K562 cells. Eur J Biochem 183:123–135

Pfeffer SR, Rothman JE (1987) Biosynthetic protein transport and sorting by the endoplasmic reticulum and Golgi. Annu Rev Biochem 56:829–852

Rose JK, Doms RW (1988) Regulation of protein export from the endoplasmic reticulum. Annu Rev Cell Biol 4:257–288

Rothman JE (1987) Protein sorting by selective retention in the endoplasmic reticulum and Golgi stack. Cell 50:521–522

Rothman JE (1981) The Golgi apparatus: two organelles in tandem. Science 213:1212–1219

Saraste J, Svensson K (1991) Distribution of the intermediate elements operating in ER to Golgi transport. J Cell Sci 100:415–430

Schutze M-P, Peterson PA, Jackson MR (1994) An N-terminal double-arginine motif maintains type II membrane proteins in the endoplasmic reticulum. EMBO J (in press)

Schweizer A, Fransen JAM, Matter K, Kreis TE, Ginsel L, Hauri H-P (1990) Identification of an intermediated compartment involved protein transport from ER to Golgi apparatus. Eur J Cell Biol 53:185–196

Signas C, Akusjarvi G, Petersson U 1986) Region E3 of human adenovirus; differences between oncogenic adenovirus-3 and the non-oncogenic adenovirus-3. Gene 50:173–184

Silberstein S, Kelleher D, Gilmor R (1992) The 48-kDa subunit of the mammalian oligosaccharyltransferase complex is homologous to the essential yeast protein WBP1. J Biol Chem 267:23658–23663

Strubin M, Long EO, Mach B (1986) Two forms of the Ia antigen-associated invariant chain result from alternative initiation at two in phase AUG's. Cell 47:619–625

Tooze SA, Tooze J, Warren G (1988) Site of addition of N-acetyl galactosamine to the E1 glycoprotein of a mouse hepatitis virus-A59. J Cell Biol 106:1475–1487

Wada I, Rindress D, Cameron P, Ou W, Doherty JJ, Louvard D, Bell AW, Dignard D, Thomas DY, Bergeron JJM (1991) SSRα and associated Calnexin are major calcium binding proteins of the endoplasmic reticulum. J Biol Chem 266:19599–19610

Wieland FT, Gleason ML, Serafini TA, Rothman JE (1987) The rate of bulk flow from the endoplasmic reticulum to the cell surface. Cell 50:289–300

Wozniak RW, Blobel G (1993) The single transmembrane domain of gp210 is sufficient for sorting to the pore membrane domain of the nuclear envelope. J Cell Biol 119:1441–1449

Protein Retention in the Golgi Stack

T. Nilsson, E. Souter, R. Watson, and G. Warren[1]

1 Introduction

The entire output of proteins newly synthesized in the endoplasmic reticulum (ER) is funneled through the Golgi stack and sorted once it reaches the *trans* Golgi network (TGN) (Griffiths and Simons 1986). Up until this point, transport occurs by default, no signals being needed for proteins to move from the ER to the Golgi and from cisterna to cisterna within the stack (Rothman and Orci 1992). This immediately raises the question of how proteins are retained along the secretory pathway; how do they resist transport to the TGN if such transport has no need of specific signals?

Two mechanisms appear to be responsible for retention along the secretory pathway. The best characterized is the retrieval mechanism which acts to recover both soluble (Pelham 1989) and membrane proteins (Jackson et al. 1993) that have been inadvertently lost from the compartment in which they normally function, or to return proteins that are part of a recycling pathway (Sweet and Pelham 1992).

Less well characterized is the retention mechanism which provides the primary means of keeping proteins in the correct compartment. Recent work on both Golgi (Swift and Machamer 1991; Nilsson et al. 1991; Munro 1991; Teasdale et al. 1992; Wong et al. 1992; Colley et al. 1992; Aoki et al. 1992; Russo et al. 1992; Tang et al. 1992; Burke et al. 1992) and ER (Wozniak and Blobel 1992; Smith and Blobel 1993) proteins shows that the retention signal lies in the membrane-spanning domain.

2 The Membrane-Spanning Domain

As an example of the evidence pointing to this domain as the retention signal, we will describe, briefly, the work we have done on the *trans* Golgi enzyme, β 1,4 galactosyltransferase (GalT) (Nilsson et al. 1991). Parts of this protein were grafted onto a reporter molecule, the human invariant chain, and the location of the hybrid proteins determined using immunofluoresence and immuno-electron microscopy, metabolic labeling and FACS analysis. The results are summarized in Fig. 1.

The human invariant chain (Ii), when transiently expressed in HeLa cells, was mostly found in the plasma membrane and early endosomes. When the cytoplasmic domain was replaced by that from GalT (GalT-I), the same results were obtained

[1] Imperial Cancer Research Fund, PO Box 123, 44 Lincoln's Inn Fields, London, WC2A 3PX, UK.

44. Colloquium Mosbach 1993
Glyco- and Cellbiology

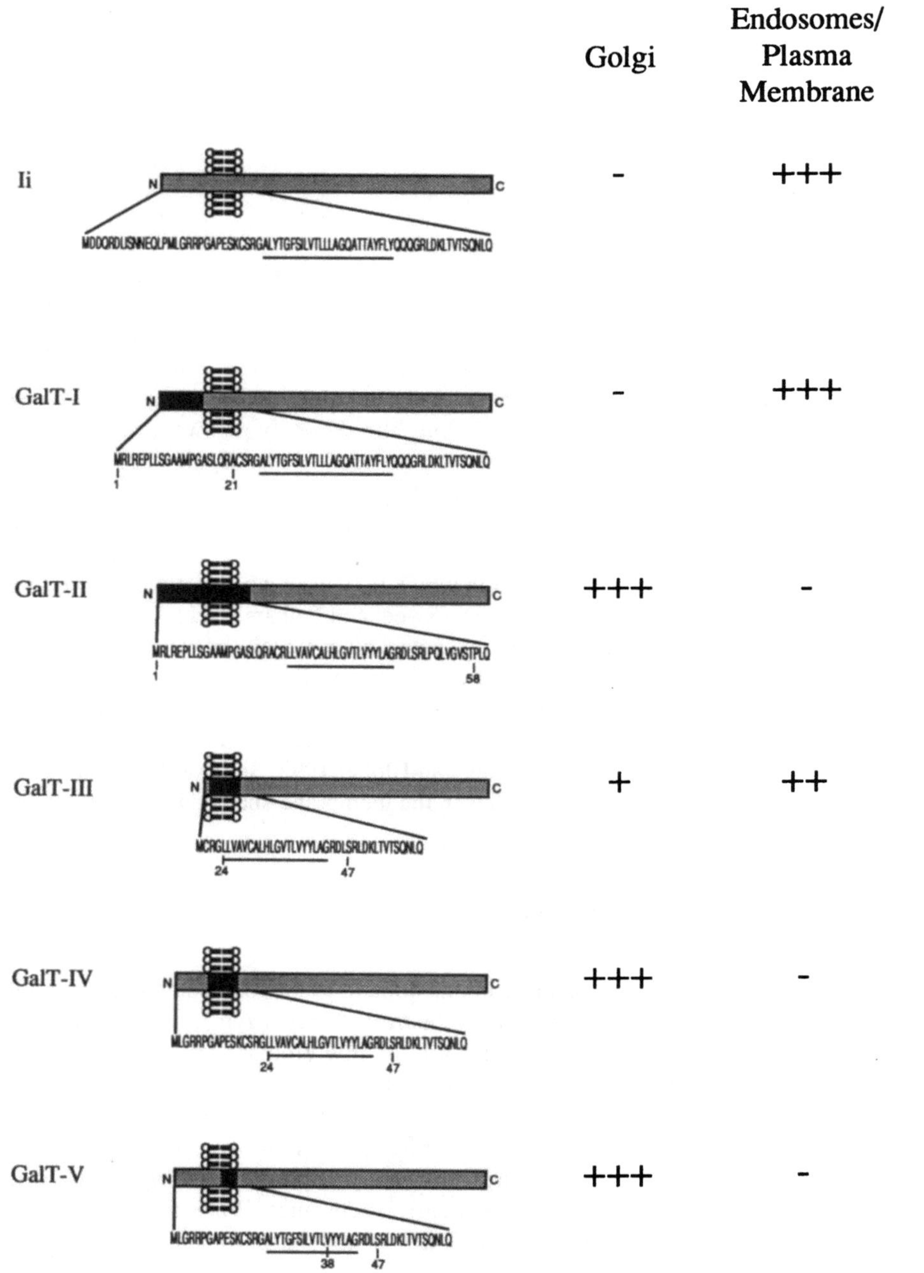

Fig. 1. Topology of the reporter molecule (Ii) and hybrid proteins containing parts of the GalT cytoplasmic and/or membrane-spanning domains. The location of these proteins in transiently transfected HeLa cells is shown *on the right*

showing that the tail alone could not specify retention in the Golgi apparatus. When the cytoplasmic and membrane-spanning domains of Ii were replaced by those from GalT (GalT-II), the hybrid proteins were now found only in the Golgi apparatus. Though these results would suggest that the membrane-spanning domain alone should specify retention, truncation of the cytoplasmic domain (GalT-III) caused some of the hybrid protein to appear on the plasma membrane. Grafting on the cytoplasmic domain of Ii (GalT-IV) restored Golgi retention, showing that it does play a role but one which can be provided by the cytoplasmic domain of the reporter molecule. Further analysis showed that only ten amino acids in the spanning region of GalT (GalT-V) were required for retention in the *trans* cisterna, the location being confirmed by immuno-electron microscopy (Nilsson et al. 1991).

Results using such transient expression systems have been confirmed using stable cell lines and, insofar as morphological studies have been carried out (Burke et al. 1992), the morphology of the Golgi apparatus is indistinguishable from that in the parental cell line. Here we report an interesting exception.

3 The 10.40 Cell Line

HeLa cells were stably transfected with the cDNA encoding the hybrid protein detailed in Fig. 2 to generate the stable cell line 10.40. The hybrid protein comprised the cytoplasmic and membrane-spanning domains of murine α1,2 mannosidase II (Mann II) together with part of the stalk region, and the lumenal domain of Ii. As expected, the hybrid protein was present in the Golgi apparatus by a number of criteria, including immunofluorescence microscopy, but the surprise came when the cells were examined by electron microscopy.

In marked contrast to the parental HeLa cell line (Fig. 3) which had typical stacks of closely apposed and flattened cisternae, the Golgi in the 10.40 cell line comprised large numbers of vesicles associated with residual cisternae which formed disorganized stacks (Fig. 4). Careful examination of the associated vesicles (Fig. 5) showed that many had the size and morphology of Golgi transport vesicles (Malhotra et al. 1989). Others had the morphology of clathrin-coated vesicles (Pearse 1987).

The explanation for this morphology is far from clear, but one intriguing possibility is a consequence of the fact that the native Ii appears to be a trimer (Marks et al. 1990). We had earlier suggested that retention occurred because the Golgi enzymes interacted with each other through their membrane-spanning domains, forming oligomers too large to enter the vesicles budding from the dilated cisternal rims

Mann II-Ii

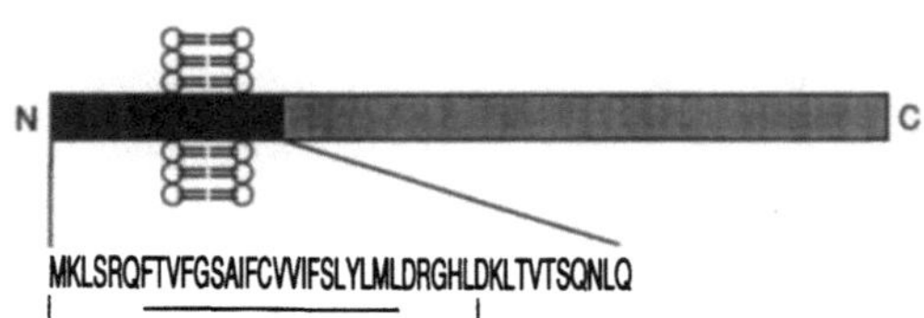

Fig. 2. Topology of the Mann II-Ii hybrid protein stably expressed by the 10.40 cell line

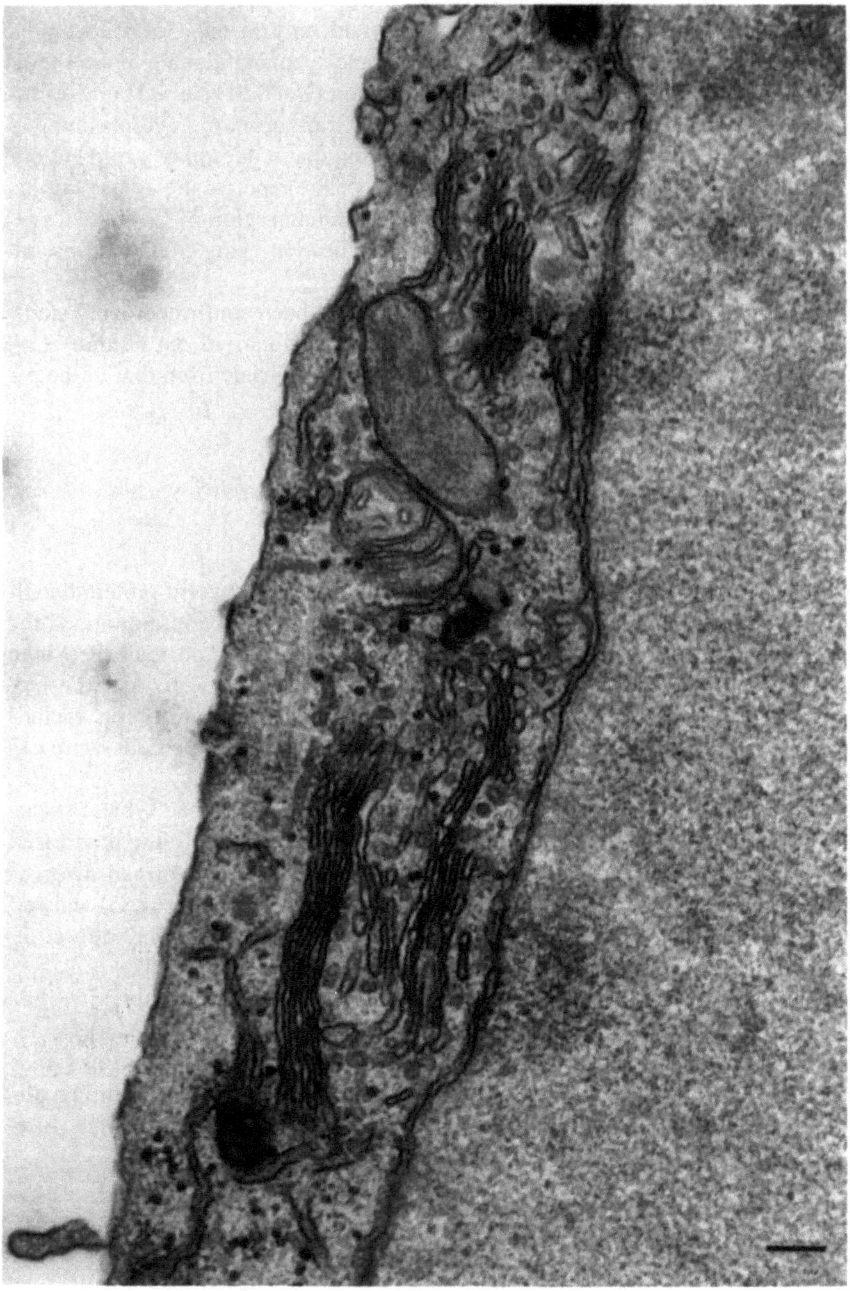

Fig. 3. Morphology of the Golgi apparatus in the parental HeLa cell line. Note the stacks of closely apposed and flattened cisternae. *Bar* 0.2 μm

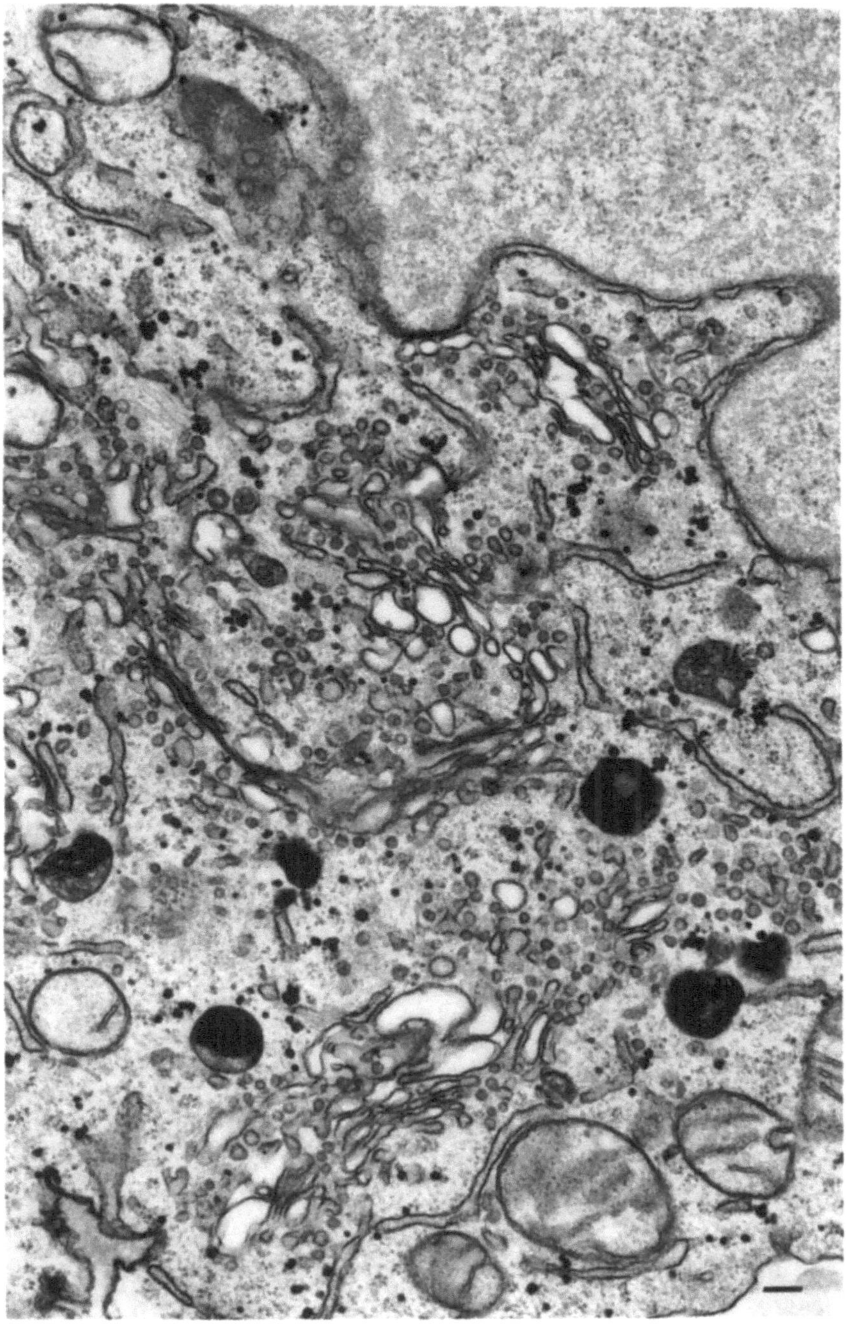

Fig. 4. Morphology of the Golgi apparatus in the 10.40 HeLa cell line. Note the large increase in the number of Golgi-associated vesicles and the disorganized nature of the stacks that remain. *Bar* 0.2 μm

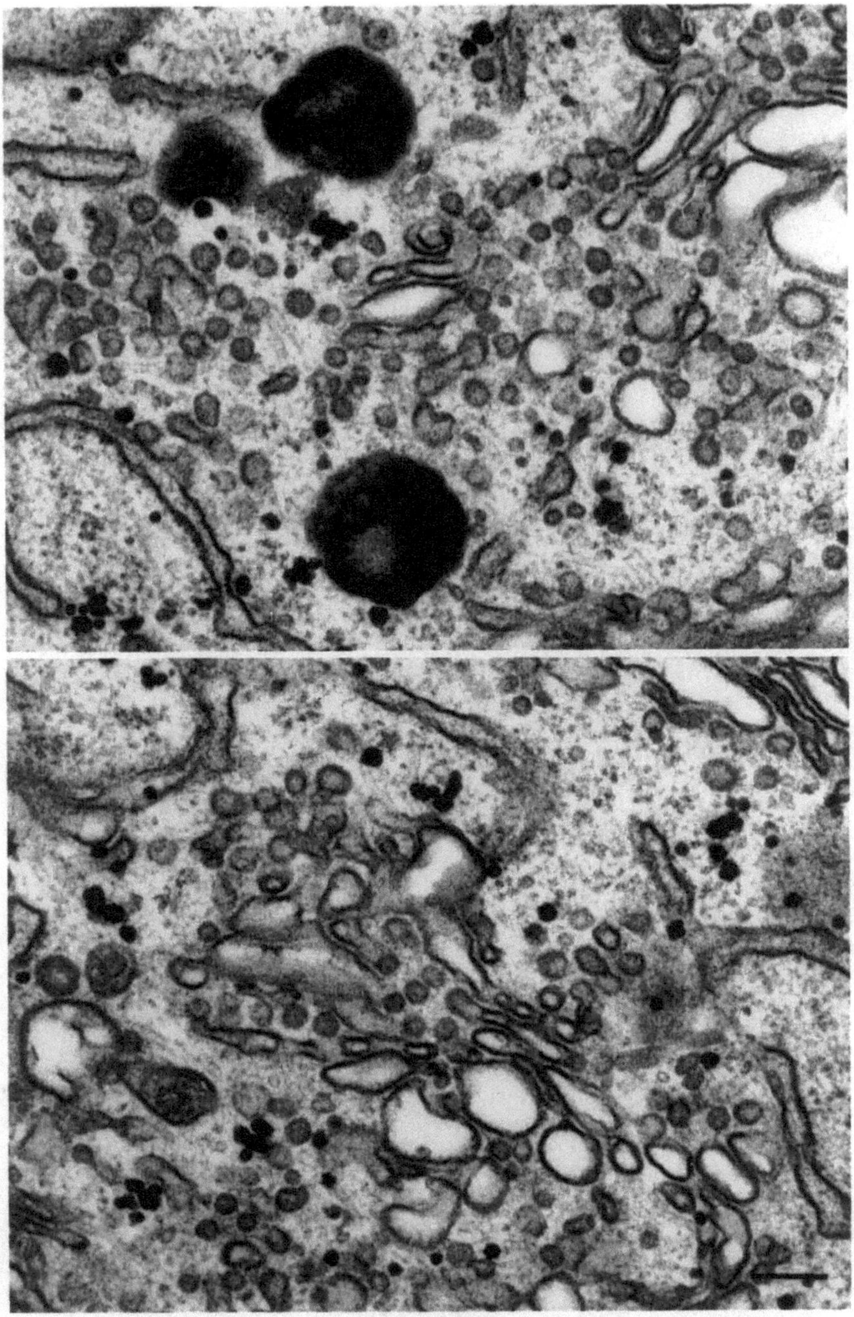

Fig. 5. Morphology of the Golgi apparatus in the 10.40 HeLa cell line. Enlargement of two of the regions in Fig. 4 showing both coated (COP and clathrin) and uncoated vesicles. *Bar* 0.2 μm

(Nilsson et al. 1991). We have recently shown that Golgi enzymes sharing the same cisterna can interact with each other (Nilsson et al. 1994). Since all Golgi enzymes so far analyzed are dimers both in vitro and in vivo (Navaratnam et al. 1988, Khatra et al. 1974; Moremen et al. 1991; Fleischer et al. 1993), this means that the oligomers would be long, linear structures. Though this would obviously aid the action of the Golgi enzymes on their substrates as they passed through the cisterna, it is difficult to see how a linear oligomer, no matter how long, could be prevented from at least partially entering the budding vesicles. We therefore suggested that these oligomers were anchored to an intercisternal matrix and we have recently obtained evidence for such a postulate (Slusarewicz et al. 1994).

If the Mann II-Ii hybrid protein is a trimer, then this could have one of two consequences. First, it could convert the linear oligomers into large, three-dimensional enzyme aggregates. This might be expected to improve retention of Golgi enzymes, but we had earlier obtained evidence without realizing it at the time that such aggregates inhibit intracellular transport. We had micro-injected mRNA encoding the monoclonal antibody 53FC3 (Burke and Warren 1984). This antibody was synthesized in the ER of micro-injected cells and transported to the Golgi apparatus, where it stopped, presumably because it was bound to Mann II. By so doing, however, it blocked transport through the Golgi apparatus of the VSV G protein, suggesting that large aggregates are incompatible with Golgi function. In the 10.40 cell line, transport is normal as shown by pulse-chase experiments following the transport of histocompatability antigens (HLA). Figure 6 compares transport to the medial and *trans* cisternae in both

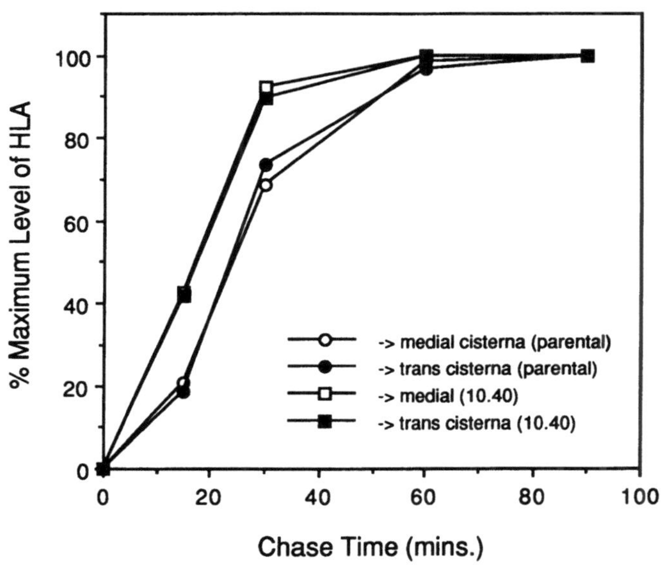

Fig. 6. Intracellular transport of HLA in the parental and 10.40 HeLa cell lines. HLA was pulsed with ^{35}S-methionine and chased for increasing times before immuneprecipitation and fractionation by SDS-PAGE. Transport to the medial cisterna was monitored by the acquisition of resistance to endoglycosidase H and to the *trans* cisterna by the acquisition of sialic acid residues detected by treatment with neuraminidase

the parental HeLa and 10.40 cell lines. If anything, transport of HLA was slightly faster in the 10.40 cell line.

The second and more likely consequence of the trimeric nature of the Mann II-Ii hybrid protein is that it somehow destabilizes the oligomers, allowing Golgi enzymes to enter the budding vesicles (Fig. 7). If this happens, there might be a mechanism to correct this error which causes these vesicles to fuse with the cisternae from which they have just budded. Such fusion events normally occur at the end of mitosis, when thousands of dispersed mitotic Golgi vesicles associate and fuse to re-form the different cisternae in the stack (Lucocq et al. 1989). In interphase cells, however, fusion with the original cisterna would constitute futile cycling, which would slow down the transport of proteins through the Golgi stack. In order to survive, the 10.40 cell line might have compensated by increasing the number of vesicles budding from each cis-

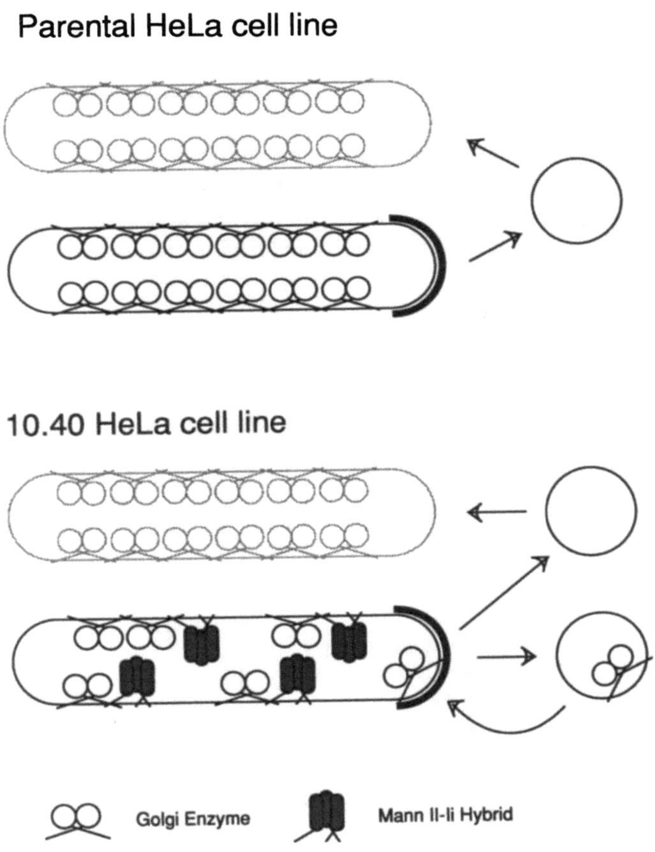

Fig. 7. model to explain the large number of vesicles seen in the 10.40 cell line. The trimeric nature of the Mann II-Ii hybrid protein is assumed to break up the oligomers that normally retain Golgi enzymes in the cisterna. Released enzymes would enter the budding vesicles, which could then no longer fuse with the next cisterna in the stack, but would fuse back with the cisterna from which they had just budded. The simplest way to restore transport to normal levels would be to increase the number of vesicles budding from each cisterna

terna. This would both restore transport to parental levels and explain the large number of transport vesicles observed.

The 10.40 cell line shows yet again that the stacked structure of Golgi cisternae is not essential for efficient transport through it. As more and more Golgi functions are uncovered, this cell line may well prove useful in showing precisely which functions do depend on an intact Golgi stack.

References

Aoki D, Lee N, Yamaguchi N, Dubois C, Fukuda MN (1992) Golgi retention of a trans-Golgi membrane protein, galactosyltransferase, requires cysteine and histidine residues within the membrane-anchoring domain. Proc Natl Acad Sci USA 89:4319–23

Baron MD, Garoff H (1990) Mannosidase II and the 125-kDa Golgi-specific antigen recognised by monoclonal antibody 53FC3 are the same dimeric protein. J Biol Chem 265:19928–19931

Burke B, Warren G (1984) Microinjection of mRNA coding for anti-Golgi antibody inhibits intracellular transport of a viral membrane protein. Cell 36:847–856

Burke J, Pettitt JM, Schachter H, Sarkar M, Gleeson PA (1992) The transmembrane and flanking sequences of β1,2-N-acetylglucosaminyltransferase I specify medial-Golgi localization. J Biol Chem 267:24433–24440

Colley KJ, Lee EU, Paulson JC (1992) The signal anchor and stem regions of the b-galactoside α2,6-sialyltransferase may each act to localize the enzyme to the Golgi apparatus. J Biol Chem 267:7784–7793

Fleischer B, McIntyre JO, Kempner ES (1993) Target size of galactosyltransferase, sialyltransferase and uridine diphosphatase in Golgi apparatus of rat liver. Biochemistry 32:2076–2081

Griffiths G, Simons K (1986) The trans Golgi network: sorting at the exit site of the Golgi complex. Science 234:438–443

Jackson MR, Nilsson T, Peterson PA (1993) Retrieval of transmembrane proteins to the endoplasmic reticulum. J Cell Biol 121:317–333

Khatra BS, Herries DG, Brew K (1974) Some kinetic properties of human-milk galactosyltransferase. Eur J Biochem 44:537–560

Lucocq JM, Berger EG, Warren G (1989) Mitotic Golgi fragments in HeLa cells and their role in the reassembly pathway. J Cell Biol 109:463–474

Malhotra V, Serafini T, Orci L, Shepherd JC, Rothman JE (1989) Purification of a novel class of coated vesicles mediating biosynthetic protein transport through the Golgi stack. Cell 58:329–336

Marks MS, Blum JS, Cresswell P (1990) Invariant chain trimers are sequestered in the rough endoplasmic reticulum in the absence of association with HLA class II antigens. J Cell Biol 111:839–855

Moremen KW, Touster O, Robbins PW (1991) Novel purification of the catalytic domain of Golgi α-mannosidase II. Characterization and comparison with the intact enzyme. J Biol Chem 266:16876–16885

Munro S (1991) Sequences within and adjacent to the transmembrane segment of α-2,6-sialylstransferase specify Golgi retention. EMBO J 10:3577–3588

Navaratnam N, Ward S, Fisher C, Kuhn NJ, Keen JN, Findlay JBC (1988) Purification, properties and cation activation of galactosyltransferases from lactating-rat mammary Golgi membranes. Eur J Biochem 171:623–629

Nilsson T, Lucocq JM, Mackay D, Warren G (1991) The membrane spanning domain of β-1,4-galactosyltransferase specifies trans Golgi retention. EMBO J 10:3567–3575

Nilsson T, How MH, Slusarewicz P, Rabouille C, Watson R, Hunte F, Watzele G, Berger EG, Warren G (1994) Kin recognition between medial Golgi enzymes in HeLa cells. EMBO J (in press)
Pearse BMF (1987) Clathrin and coated vesicles. EMBO J 6:2507–2512
Pelham HR (1989) Control of protein exit from the endoplasmic reticulum. Ann Rev Cell Biol 5:1–23
Rothman JE, Orci L (1992) Molecular dissection of the secretory pathway. Nature 355:409–416
Russo RN, Shaper NL, Taatjes DJ, Shaper JH (1992) β1,4-galactosyltransferase: a short NH_2-terminal fragment that includes the cytoplasmic and transmembrane domain is sufficient for Golgi retention. J Biol Chem 267:9241–9247
Slusarewicz P, Nilsson T, Hui N, Watson R, Warren G (1994) Isolation of a intercisternal matrix that binds medial Golgi enzymes. J Cell Biol (in press)
Smith S, Blobel G (1993) The first membrane spanning region of the lamin B receptor is sufficient for sorting to the inner nuclear membrane. J Cell Biol 120:631–637
Sweet DJ, Pelham HRB (1992) The *S. cerevisiae* SEC20 gene encodes a membrane glycoprotein which is sorted by the -HDEL retrieval system. EMBO J 11:423–432
Swift AM, Machamer CE (1991) A Golgi retention signal in a membrane-spanning domain of coronavirus-E1 protein. J Cell Biol 115:19–30
Tang BL, Wong SH, Low SH, Hong W (1992) The transmembrane domain of N-glucosaminyltransferase I contains a Golgi retention signal. J Biol Chem 267:10122–10126
Teasdale RD, D'Agostaro G, Gleeson PA (1992) The signal for Golgi retention of bovine β1,4-galactosyltransferase is in the transmembrane domain. J Biol Chem 267:4084–4096
Wong SH, Low SH, Hong W (1992) The 17-residue transmembrane domain of the β-galactoside α2,6-sialyltransferase is sufficient for Golgi retention. J Cell Biol 117:245–258
Wozniak RW, Blobel G (1992) The single transmembrane segment of gp210 is sufficient for sorting to the pore membrane domain of the nuclear envelope. J Cell Biol 119:1441–1449

Signal-Mediated Targeting of Lysosomal Membrane Glycoproteins

K. von Figura, A. Hille-Rehfeld, L. Lehmann, C. Peters, and V. Prill[1]

1 Introduction

Lysosomes are the major digestive organelles of eukaryotic cells. For this purpose they are equipped with a cohort of some 40 to 50 acid hydrolases that degrade a wide spectrum of macromolecules including proteins, lipids, nucleic acids, and polysaccharides into their monomeric constituents. The bulk of these acid hydrolases and of their cofactors are soluble components of the lysosomal matrix. The lysosomal membrane separates the potentially hostile lysosomal hydrolases from the rest of the cellular compartments and is responsible for acidification of the lumen. Furthermore, the lysosomal membrane mediates the transport of degradation products such as amino acids, monosaccharides, nucleosides, and ions from the lysosomal lumen to the cytoplasm. Moreover, the membrane regulates the fusion and fission events between lysosomes and other organelles.

The physiological function of lysosomes depends on the continuous supply with newly synthesized components of the lysosomal matrix and lysosomal membrane. In the past, the biosynthesis and transport of the soluble matrix proteins has been widely studied, mostly using acid hydrolases as models. Mannose 6-phosphate receptors have been shown to be critical for the targeting of soluble lysosomal proteins. The biosynthesis of the mannose 6-phosphate residues on lysosomal proteins and their targeting to lysosomes has been the subject of several recent reviews (Kornfeld 1992; Kornfeld and Mellmann 1989; von Figura and Hasilik 1986). Briefly, a proteinaceous determinant in newly synthesized lysosomal proteins serves as a recognition determinant for a membrane-associated transferase located in the *cis* Golgi reticulum which transfers N-acetylglucosamine-1-phosphate from UDP-GlcNAc to the C6 hydroxyl-group of mannose residues in high-mannose oligosaccharides. By comparing the sequences of the more than 20 soluble lysosomal proteins for which the cDNAs are known today, it has not been possible to identify the proteinaceous determinant(s), which must be common to all soluble lysosomal proteins and distinguish them from the multitude of secretory and membrane-associated glycoproteins passing the *cis*-Golgi reticulum. For the bilobal cathepsin D precursor two noncontiguous sequences in the carboxyl lobe were identified as major determinants of the phosphotransferase recognition domain. However, also several amino lobe determinants contribute to the recognition by the phosphotransferase and it is at present not decided, whether all these noncontiguous sequences are part of one phosphotransferase recognition do-

[1] Georg-August-Universität, Abt. Biochemie II, Gosslerstr. 12d, D-37073 Göttingen, FRG.

44. Colloquium Mosbach 1993
Glyco- and Cellbiology

main or whether cathepsin D precursors contain two independent recognition sites (Baranski et al. 1990, 1992).

Removal of the outer N-acetylglucosamine group by a Golgi-associated α-N-acetylglucosaminidase generates the mannose 6-phosphate residues, which serve as a recognition marker for the binding to mannose 6-phosphate receptors in the *trans*-Golgi reticulum. The receptor-lysosomal protein complexes are packaged into clathrin-coated vesicles that fuse with membranes of the endosomal compartment. Early endosomes appear to be among the accepting sites for these vesicles (Ludwig et al. 1991). Within the endosomal compartment, the receptor-lysosomal protein complexes are dissociated by the low pH. While the receptors recycle back to the *trans* Golgi reticulum or the plasma membrane, the lysosomal proteins are delivered to lysosomes. At present, there is no consensus as to whether (late) endosomes and lysosomes are two separate organelles that communicate via transport vesicles or whether there is a gradual maturation of (late) endosomes to lysosomes by processes of continuous fusion and fission of vesicles.

A small fraction of newly synthesized lysosomal proteins is delivered to the extracellular milieu. This has been thought to reflect missorting and is ascribed to an inefficiency of the receptor-dependent retention mechanism. Recently, however, it has become apparent that part of the initially receptor-bound lysosomal proteins end up in the secretions, suggesting that unphysiological secretion of lysosomal enzymes involves mannose 6-phosphate receptors (Chao et al. 1990). Whatever the mechanisms may be by which newly synthesized lysosomal enzymes appear in the secretions, they are capable of binding to mannose 6-phosphate receptors that are exposed at the cell surface and which become rapidly internalized via clathrin-coated vesicles. This retrieval mechanism ensures that also transiently secreted lysosomal proteins will end up in lysosomes, provided they contain mannose 6-phosphate residues.

Interest in the transport of lysosomal membrane proteins developed much later. Significant progress has been made in recent years in unraveling the targeting mechanisms by which this class of proteins is sorted to its final destination. Using lysosomal membrane glycoproteins, the function of which is unknown (designated with various acronyms, of which lamp, lysosome-associated membrane protein, and lgp, lysosomal membrane glycoprotein, are the most widely used), and lysosomal acid phosphatase (LAP) as models, it has been shown that sorting of these membrane proteins is independent of mannose 6-phosphate receptors and depends on short tyrosine-containing sequences within their cytoplasmic tails, which are necessary and sufficient to target them to lysosomes (for a recent review see Fukuda 1991). This chapter will focus on the signals mediating the transport of newly synthesized LAP in nonpolarized and polarized cells to lysosomes, and the cytoplasmic receptors recognizing these signals.

2 Pathway of LAP in Nonpolarized and Polarized Cells

The earliest indication that trafficking of LAP might differ from that of other lysosomal enzymes came from the observation that I-cell fibroblasts contain normal levels of LAP. Due to the deficiency of the phosphotransferase, a marked deficiency of lysosomal enzymes whose targeting depends on mannose 6-phosphate receptors is characteristic of I-cell fibroblasts. The normal LAP activity in phosphotransferase-deficient fibroblasts was particularly intriguing, since the two other lysosomal enzymes that were found to be normal in I-cell fibroblasts (glucocerebrosidase and a heparan-sulfate-specific N-acetyltransferase) are membrane-associated, while the bulk of LAP is soluble, similar to many lysosomal enzymes that are deficient in I-cell fibroblasts. Initial attempts to characterize the biosynthesis and transport of LAP in fibroblasts suggested that it is synthesized and transported as a mannose 6-phosphate containing polypeptide (Lemansky et al. 1985; Waheed and van Etten 1989). In retrospect, these studies were misled by a specific cross-reactivity of the LAP antisera with a mannose 6-phosphate-containing polypeptide unrelated to LAP (Gottschalk 1990).

Progress in the understanding of the transport of LAP was promoted by the cloning and expression of its cDNA. The cDNA that encodes human LAP predicted a polypeptide with a cleavable signal peptide at the N-terminus, a large luminal domain of 348 residues, a single hydrophobic transmembrane domain, and a small C-terminal domain of 19 residues exposed at the cytoplasmic face of the membrane (Fig. 1) (Pohlmann et al. 1988). In vitro translation in the presence of microsomal membranes showed that LAP was indeed synthesized as a type I membrane protein. Expression of the LAP cDNA in baby hamster kidney (BHK) cells greatly facilitated the analysis of the biosynthesis, posttranslational modification, and transport of the LAP (Fig. 2). LAP is synthesized as a membrane-associated precursor, in which all eight potential N-glycosylation sites become glycosylated. Transfer of the precursor to the *trans*-Golgi is associated with processing of most of the oligosaccharides to complex type structures, and requires about 0.5 h in BHK cells. Much later (half-time about 13–14 h), proteolytic processing of LAP is observed (Waheed et al. 1988). This proteolytic processing is assumed to be restricted to lysosomes, and was shown to involve the removal of the cytoplasmic tail and the release of the luminal domain of LAP into the lysosomal matrix (Gottschalk et al. 1989). The long period elapsing between passage of the *trans*-Golgi and proteolytic processing was shown to be due to a slow delivery of LAP precursors to dense lysosomes (half-time about 6–7 h) followed by a slow proteolytic processing within lysosomes (half-time in lysosomes about 6–7 h). The release of the soluble form of LAP from the membrane depends on

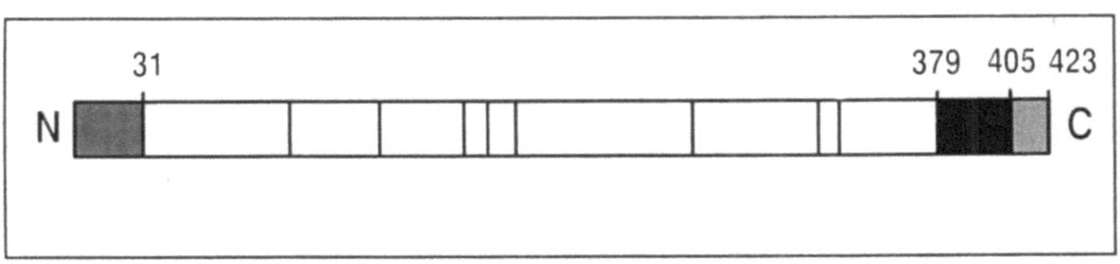

Fig. 1. *Predicted structure of the LAP precursor.* The *stippled areas* represent the N-terminal signal peptide and the C-terminal cytoplasmic tail, the *dark area* the transmembrane domain. *Vertical lines in the open box* indicate the position of the eight N-glycosylation sites in the luminal domain. *Numbers* give the position of the aminoterminus of mature LAP, the first residue of the transmembrane and cytoplasmic domains, and of the carboxyterminus of the LAP precursor

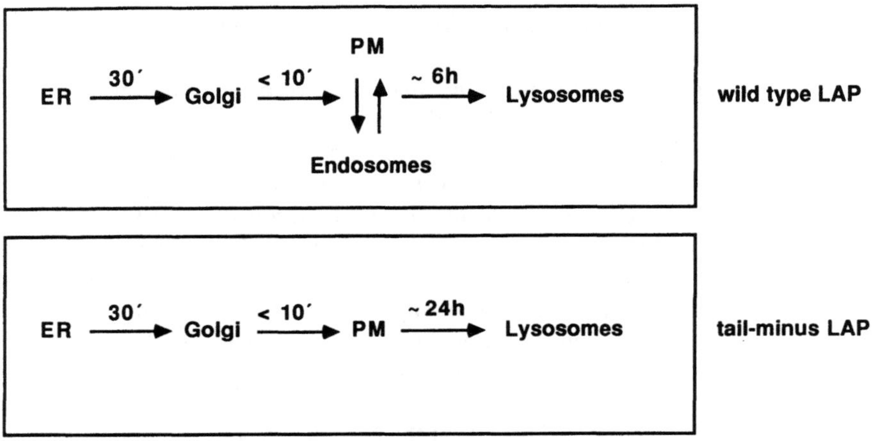

Fig. 2. *Biosynthetic pathway of wild-type LAP and the tail-minus form of LAP. Numbers above the arrows* give the half-time for the transport

an aspartylproteinase, which is likely to be cathepsin D (Gottschalk et al. 1989; Himeno et al. 1991).

The slow transfer of LAP precursor from the *trans*-Golgi to dense lysosomes can be explained by a repeated recycling of LAP between endosomes and the plasma membrane. In BHK cells, LAP remains on average for about 5–6 h in this endosome/plasma membrane pool. At steady state, about 20% of LAP precursors in this pool reside at the plasma membrane and 80% in internal membranes. The times for one cycle were calculated to be between 6 and 20 min. Thus, a single LAP precursor recycles in BHK cells 15 to 50 times between endosomes and the plasma membrane before it is delivered to dense lysosomes (Braun et al. 1989).

A matter of dispute is whether LAP precursors are delivered from the *trans*-Golgi reticulum first to endosomes and reach the cell surface from there or vice versa. The half-time for transport of LAP from the *trans*-Golgi reticulum to the cell surface is ≤ 10 min in BHK cells. This is comparable with the transit time that has been measured for the delivery of many secretory proteins from the *trans*-Golgi to the plasma membrane. Furthermore, the *trans*-Golgi to plasma membrane transit time of a soluble mutant of LAP, which lacks the transmembrane and the cytoplasmic domains, was indistinguishable from that found for the membrane-associated precursor (M. Braun unpubl.). These observations are compatible with the view that LAP is transported directly from the TGN to the plasma membrane. However, these kinetic criteria cannot exclude an indirect route to the cell surface via endosomes. The resolution of this question will depend on the isolation of vesicles that bud off from the *trans*-Golgi reticulum, and on the determination of the association of LAP with vesicle subpopulations enriched in secretory proteins or in proteins with endosomal destination, e.g., mannose 6-phosphate receptors and their ligands.

The amout of LAP at the cell surface may vary. While in BHK cells more than 10% of the LAP precursor was found at the cell surface, it was below the level of biochemical detection in the plasma membrane of rat liver hepatocytes (Tanaka et al.

1990). Immunocytochemical staining revealed plasma membrane-associated LAP in rat liver hepatocytes (Yokota et al. 1989), human skin fibroblasts (Parenti et al. 1987), and exocrine pancreatic cells (J. Tooze pers. comm.). Analysis of LAP transport in transfected MDCK cells, a cell line derived from canine kidney, lent further support to the notion that transport of LAP to lysosomes includes passage of the plasma membrane. In MDCK cells, less that 1% of the membrane-associated LAP is present at the cell surface. When MDCK cells are grown in a polarized fashion, surface-associated LAP is readily detectable at the basolateral and barely detectable at the apical domain of the plasma membrane. Analysis of the transport of newly synthesized LAP precursors in polarized MDCK cells showed that the precursors are directly sorted from the *trans*-Golgi reticulum to the basolateral domain (Prill et al. 1993).

It became readily apparent that the mechanisms by which LAP is targeted to lysosomes differ fundamentally from the mannose 6-phosphate receptor-dependent targeting of soluble lysosomal proteins. Thus, at no period of their life cycle do LAP precursors contain mannose 6-phosphate residues. Moreover, targeting of LAP is independent of N-glycosylation and is resistant to ammonium chloride (Gottschalk et al. 1989b). Since the latter dissipates pH gradients between acidic organelles and the cytoplasm, an acid pH-dependent step is not critical for targeting of LAP. Experiments described in the following section revealed that transport of LAP in nonpolarized and polarized cells critically depends on sorting information provided by its cytoplasmic tail.

3 Signals Mediating the Trafficking of LAP

The luminal domain of LAP is highly homologous to prostatic acid phosphatase, which is a secretory protein (Peters et al. 1989). This suggested that potential trafficking signals in the luminal domain are not sufficient for targeting LAP to lysosomes and that critical signals may reside in the transmembrane and/or cytoplasmic domain of LAP. This was further supported by the observation that a truncated LAP mutant lacking the transmembrane and cytoplasmic domains is secreted.

The presence of a signal for rapid internalization within the cytoplasmic tail of LAP became apparent when the transport of a mutant LAP lacking the cytoplasmic tail was analyzed. This tail-minus form of LAP was transported from the endoplasmic reticulum to the *trans*-Golgi and from there to the cell surface with kinetics indistinguishable from those observed for wild-type LAP. The tail-minus form, however, accumulated at the cell surface. The rate of internalization decreased from 11%/min found for wild-type LAP to 1.6%/min. The transient arrest at the cell surface prolonged the half time for transport of newly synthesized LAP to dense lysosomes from 6–7 h to about 24 h (Fig. 2). The cytoplasmic tail of LAP was fused to the luminal and transmembrane domain of hemagglutinin. The wild-type form of hemagglutinin with a cytoplasmic tail of 11 residues behaves in BHK cells as a resident plasma membrane protein. The fusion protein with the cytoplasmic tail of LAP was transported to the cell surface, from where it was rapidly internalized. This proved that the cytoplasmic tail of LAP contains a signal for rapid internalization which is sufficient to induce internalization of a reporter molecule (Peters et al. 1990).

The structural requirements for rapid internalization were characterized by analyzing the internalization in BHK cells of two series of mutants, in which either the cytoplasmic tail of LAP was progressively shortened starting from the C-terminus, or individual amino acids were replaced by alanine or more related amino acids (Lehmann et al. 1992). The truncation mutants provided the information that the aminoterminal 12 residues are sufficient for rapid internalization and thus contain the internalization signal. Interestingly, partial loss of up to five of the C-terminal residues was also associated with a substantial decrease of the internalization rate. This was ascribed to a modifying effect of the C-terminal portion of the tail, which can affect the exposure or conformation of the internalization signal residing within the aminoterminal 12 residues. Substitution of individual residues by alanine identified among the aminoterminal 12 residues the sequence of 411 Pro-Gly-Tyr-Arg-His-Val 416 as critical for internalization. Replacing Gly 412, Tyr 413, Arg 414, or Val 416 by alanine decreased the internalization to values observed for the tail-minus form of LAP and that of Pro 411 and His 414 reduced the internalization rate to about half normal values (Fig. 3). Conservative substitution of Tyr 413 by phenylalanine resulted in a loss of the internalization signal, suggesting that the hydroxyl group of the tyrosine is critical for the conformation of the internalization signal or an essential part of its recognition domain.

A characteristic feature of the biosynthetic pathway of LAP is the repeated cycling of LAP between the endosomes and plasma membrane before it is delivered to lysosomes. At present it is not clear whether the cycling back to the cell surface and/or the

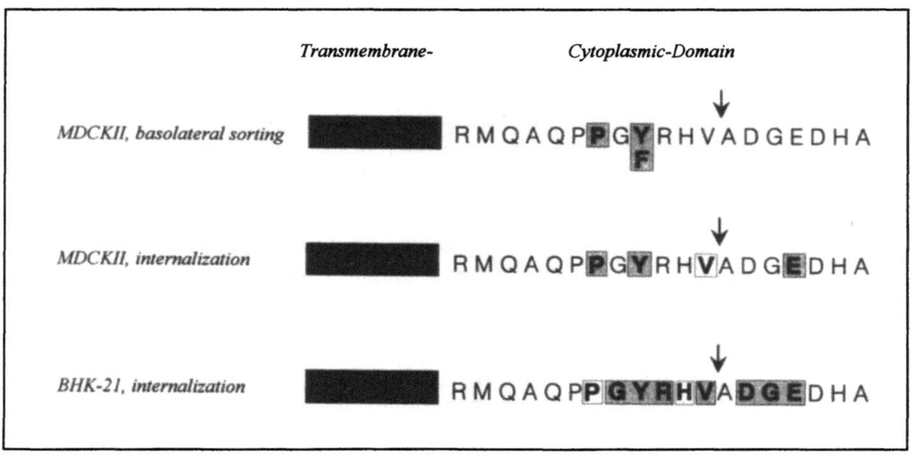

Fig. 3. *Transport signals in the cytoplasmic tail of LAP.* Indicated is the transmembrane domain (*filled box*) and the sequence of the cytoplasmic tail. *Bold letters in shaded boxes* indicate residues that cannot be replaced by alanine without severe loss of the transport signal. *Bold letters in open boxes* indicate residues whose substitution by alanine causes a partial loss of the transport signal. Other residues can be replaced by alanine without alterning the signal activity. The aminoterminal four residues (RMQA) were not examined by alanine substitution. The *arrow* indicates the C-terminal border of the transport signal identified by truncation mutants

transport to dense lysosomes depend(s) on a specific signal. It is conceivable that properties of the luminal domain and/or the quaternary structure are critical for these transport steps.

In MDCK cells grown in a polarized fashion, newly synthesized LAP is transported from the *trans*-Golgi reticulum selectively to the basolateral domain of the plasma membrane (see above). To examine whether sorting to the basolateral membrane depends on a signal, the tail-minus and the secretory form of LAP were expressed in MDCK cells. In contrast to wild-type LAP, the tail-minus form was transported to the apical plasma membrane domain and the secretory form of LAP accumulated in the secretions above the apical membrane. The most likely explanation is the presence of a basolateral sorting signal in the cytoplasmic tail of LAP. However, the possibility cannot be excluded that LAP contains an apical sorting signal within its luminal domain that is masked by the cytoplasmic tail, and that transport of wild-type LAP to the basolateral membrane does not depend on a signal. To differentiate between these two possibilities, the tail of hemagglutinin was replaced by that of LAP. Wild-type hemagglutinin is sorted to the apical membrane. The presence of the LAP tail redirected that chimeric hemagglutinin to the basolateral membrane, showing that the cytoplasmic tail contains a basolateral sorting signal (Prill et al. 1993).

The structural requirements for the basolateral sorting signal were characterized (Prill et al. 1993) using the same series of truncation and substitution mutants used for characterization of the internalization signal. For basolateral sorting, the aminoterminal 12 residues of the tail were sufficient. Further shortening of the tail abolished the basolateral sorting. Thus, the basolateral sorting signal is located in the same region of the cytoplasmic tail as the signal for rapid internalization. Substitution of individual residues by alanine identified Pro 411 and Tyr 413 as the essential residues within the aminoterminal 12 residues, while the conservative exchange of Tyr 413 by phenylalanine or that of Gly 412, Arg 414, and Val 416 by alanine did not affect basolateral sorting (Fig. 3). Since the latter abolished rapid internalization in BHK cells, this suggested that the structural requirements for the internalization signal are much more stringent than for basolateral sorting.

Since it was conceivable that different requirements for basolateral sorting and rapid internalization of LAP were due to the use of different cell types and/or species, we analyzed also the internalization of LAP in MDCK cells (Prill et al. 1993). This analysis showed that only LAP mutants with an intact basolateral sorting signal were internalized. As expected from the higher stringency requirements for rapid internalization, some of the mutants which are sorted correctly to the basolateral plasma membrane had lost their internalization signal. This was true for the conservative substitution of Tyr 413 by phenylalanine and the substitution of Val 416 by alanine (Fig. 3). Thus, the requirements for basolateral sorting and rapid internalization show also clear differences if examined in one cell type. However, different requirements were also found between the internalization of LAP in BHK and MDCK cells. For example, the substitution of Gly 412 or Arg 414 by alanine severely decreases internalization in BHK but not in MDCK cells. These differences have to be ascribed to cell type- or species-dependent differences of the cytoplasmic receptors for the signals in BHK and MDCK cells. These data clearly show that the signals for basolateral sorting and rapid internalization of LAP overlap or are even represented by the identical sequence and are recognized at the *trans*-Golgi network and the plasma membrane by different receptors with specific and distinct structural requirements for binding.

The dominant function of the cytoplasmic tail of LAP in mediating basolateral sorting and rapid internalization of reporter molecules indicated that the tail folds independently of the luminal and transmembrane domains. We therefore reasoned that an in vitro synthesized peptide corresponding to the cytoplasmic LAP tail may fold in solution and adopt a conformation sufficiently stable to be analyzed by 2D-NMR. This assumption held true and it was shown that the Pro-Pro-Gly-Tyr tetrapeptide corresponding to residues 410–413 in LAP forms a type I β–turn (Eberle et al. 1991). At any given time, ≥ 50% of the total peptide exhibits the turn. N- and C-terminal of the turn nascent helices were observed. Replacing in the peptide the tyrosine by phenylalanine or alanine decreases the probability of a β–turn conformation by 25 and 50%, respectively (Lehmann et al. 1992). It should be noted that the substitution of Tyr 413 by phenylalanine abolishes the signal for internalization but not for basolateral sorting, while substitution by alanine destroys both signals. Thus, the cytoplasmic receptors mediating basolateral sorting or rapid internalization both require a β–turn, whereas that for basolateral sorting might tolerate a higher instability of the turn.

Comparison of the structural information provided by the 2D-NMR analysis with the functional characterization of transport signals indicated that the hexapeptide Pro-Gly-Tyr-Arg-His-Val which harbors the internalization and basolateral sorting signals is made up of a tight turn and a nacent helix. The three aminoterminal residues contribute to the tight turn, and the three carboxyterminal residues contribute to the nascent helix.

4 Receptors for Transport Signals in LAP

The rapid internalization of LAP and the similarity of its internalization signal with that of other membrane receptors internalized through clathrin-coated vesicles suggested that LAP is internalized via clathrin-coated pits. Immunogold labeling showed that at various stages of invaginating coated pits, LAP colocalized with clathrin and the plasma membrane-specific adaptor HA-2. Quantitation of the immunogold label, however, showed an even distribution of LAP in coated and in noncoated areas of the plasma membrane. Since the tail-minus form of LAP was partly excluded from coated areas, this indicated a localization of LAP in clathrin-coated membrane areas dependent on the cytoplasmic tail. The transient inhibition of the internalization of LAP after microinjection of monoclonal antibodies against clathrin or HA-2 provided further evidence for the role of clathrin and HA-2 for internalization of LAP (Hille et al. 1992). The plasma membrane adaptor HA-2, as well as the closely related adaptor HA-1, found at the *trans*-Golgi reticulum are heterotetrameric complexes of two 100 kDa subunits and two smaller subunits of about 50 and 20 kDa. The 100 kDa α- and β-subunit of HA-2 adapters (α- and β-adaptins) are homologous to the γ- and the β'-subunits (γ- and β'-adaptins), respecitively, of HA-1 adaptors. Both adapters function in the concentration of membrane proteins in clathrin-coated membrane areas at the *trans*-Golgi reticulum (HA-1) or the plasma membrane (HA-2). HA-2 binds to tyrosine-containing internalization signals in membrane proteins, promotes the assembly of the clathrin coat, and thereby concentrates the membrane proteins in clathrin-coated pits (for review see Pearse and Robinson 1990). The β-subunit of HA-2 adaptors has been shown to bind both clathrin (Ahle and Ungewickell 1989) and the cytoplasmic domain of the asialoglycoprotein receptor, which processes a tyrosine-containing internalization signal (Beltzer and Spiess 1991). To examine the binding of LAP-cytoplasmic tail to HA-2 adaptors, we immobilized peptides representing the wild-type or mutant forms of cytoplasmic tail of LAP to Affi-Gel 10. A mixture of HA-2 and HA-1, as well as purified adapters or clathrin were passed over the peptide-Affi-Gel 10 matrix (Fig. 4) (Sosa et al. 1993). Matrices substituted with the LAP tail peptide or a truncated form containing the aminoterminal 12 residues retained more than 90% of the HA-2 adapters, while less than 20% of the HA-1 adapters were bound. The binding of the HA-2 adapters was saturable. When a mutant tail peptide, in which the tyrosine was replaced by alanine, was used as affinity matrix, binding of HA-2 adapters dropped to values found for HA-1. The poor binding of HA-1 adapters to the LAP tail peptide was specific, since the same adaptors bound well to the tail of the M_r 46 000 mannose 6-phosphate receptor. This receptor is sorted at the *trans*-Golgi reticulum into clathrin-coated vesicles and therefore assumed to contain a binding site for HA-1. A discrepancy between the in vitro binding of HA-2 adaptors and the internalization in vivo was found for the substitution of the tyrosine residue in the LAP tail by phenylalanine. While this substitution abolishes rapid internalization of LAP in BHK and MDCK cells, HA-2 binds to a peptide with this substitution with a comparable affinity as to the wild-type tail of LAP. The 2D-NMR analysis of the phenylalanine-containing tail peptide revealed that this substitution reduces the tendency to form a β-turn by 25%. Apparently, the remaining turn-forming capacity is sufficient for in vitro but not for in vivo binding to HA-2 adapters. When fragmented HA-2 adapters, of which the carboxyterminal head part was cleaved off from the aminoterminal trunk of the 100-kDa adaptins by partial digestion with trypsin, were subjected to affinity chromatography, the trunks were recovered in the bound and the heads in the flow through fraction. Since trunks are also sufficient for binding to clathrin, this would suggest that the aminoterminal trunk of the β-adaptin contains the binding sites for clathrin (Schröder and Ungewickell 1991) and tyrosine containing internalization signals.

The cytoplasmic receptors for the basolateral sorting signal remain to be detected. They may be related to the HA-1 and HA-2 adapters. However, our knowledge on the formation of basolateral vesicles at the *trans*-Golgi reticulum and involvement of coat proteins is as yet too fragmentary to conclude that their formation is analogous to that of clathrin-coated vesicles.

5 Concluding Remarks

In the cytoplasmic tail of LAP, signals have been identified for rapid internalization and basolateral sorting. Both signals reside in the aminoterminal 12 residues of the cytoplasmic tail and the essential residues are found within the hexapeptide Pro-Gly-

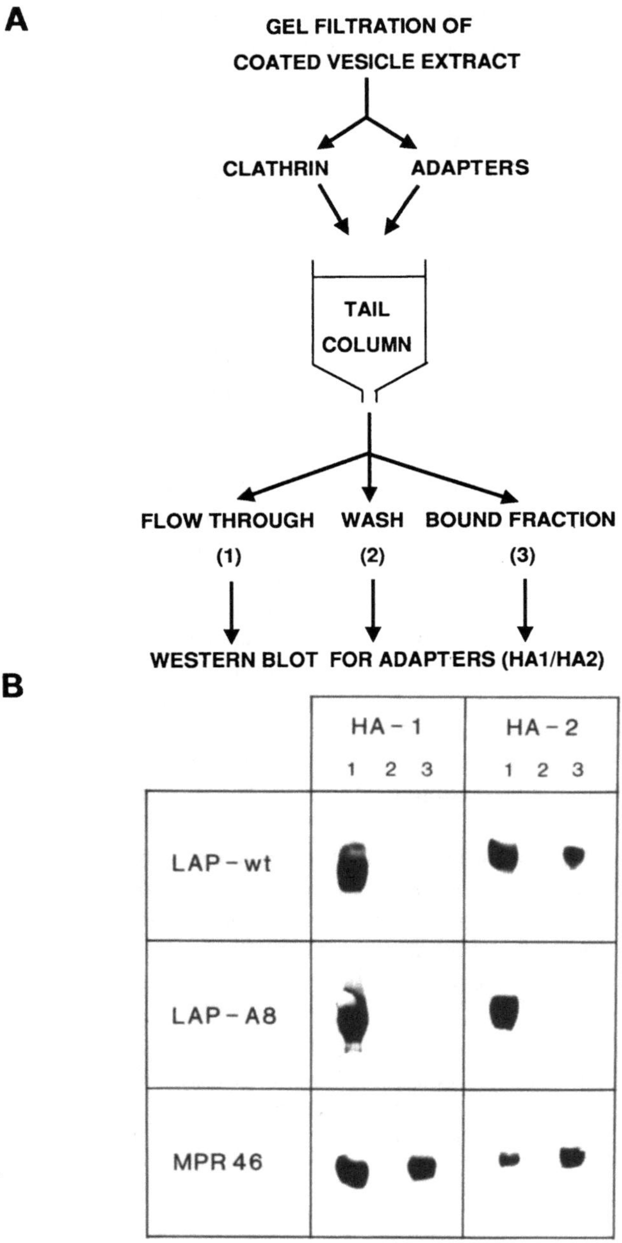

Fig. 4 A, B. *In vitro binding of plasma membrane-coated vesicle adapters to the cytoplasmic domain of LAP.* **A** Experimental procedures for purification and affinity chromatography of adapters from coated vesicles of bovine brain. **B** Results of adapter binding to tail columns analyzed by Western blotting for Golgi (*HA-1*) or plasma membrane (*HA-2*) adapters. *LAP-wt* Peptide corresponding to the cytoplasmic tail of wild-type LAP; *LAP-A8* modified peptide with tyrosine 8 replaced by alanine: *MPR46* peptide corresponding to the cytoplasmic tail of M_r 46 000 mannose 6-phosphate receptor

Tyr-Arg-His-Val. The hexapeptide is part of a β-turn and a nascent helix. The overlap of the two signals suggests that the same structural motif in the cytoplasmic tail of LAP is recognized by the receptors at the *trans*-Golgi reticulum and plasma membrane, which mediate basolateral sorting and internalization, respectively. The two receptors, however, have distinct requirements for binding and hence tolerate mutations distinctly.

The productive interactions of one (multifunctional) structural motif with different receptors located at different sites bears formal analogy to the opening of different doors with different locks by one master key. Multifunctional sorting signals may be a more common principle in cell biology. For example, in the cytoplasmic tail of the M_r 46 000 mannose 6-phosphate receptor, a Leu-Leu-containing motif mediates sorting at the *trans* Golgi reticulum (Johnson and Kornfeld 1992) and rapid internalization (B. Weber and R. Pohlmann unpubl.).

Acknowledgments. Work done in the authors' laboratory was supported by the Deutsche Forschungsgemeinschaft and the Fonds der Chemischen Industrie.

References

Ahle S, Ungewickell E (1989) Identification of a clathrin-binding subunit in the HA-2 adaptor protein complex. J Biol Chem 264:20089–20093

Baranski TJ, Faust PL, Kornfeld S (1990) Generation of a lysosomal enzyme targeting signal in the secretory protein pepsinogen. Cell 63:281–291

Baranski TJ, Cantor AB, Kornfeld S (1992) Lysosomal enzyme phosphorylation I. Protein recognition determinants in both lobes of procathepsin D mediate its interaction with UDP-GlcNAc, lysosomal enzyme N-acetylglucosamine-1-phosphotransferase. J Biol Chem 267:23342–23348

Beltzer JP, Spiess M (1991) In vitro binding of the asialoglycoprotein receptor to the beta adaptin of plasma membrane-coated vesicles. EMBO J 10:3735–3742

Braun M, Waheed A, von Figura K (1989) Lysosomal acid phosphatase is transported to lysosomes via the cell surface. EMBO J 8:3633–3640

Chao HHJ, Waheed A, Pohlmann R, Hille A, von Figura K (1990) Mannose 6-phosphate receptor-dependent secretion of lysosomal enzymes. EMBO J 9:3507–3513

Eberle W, Sander C, Klaus W, Schmidt B, von Figura K, Peters C (1991) The essential tyrosine of the internalization signal in lysosomal acid phosphatase is part of a β-turn. Cell 67:1–20

Fukuda M (1991) Lysosomal membrane glycoproteins. J Biol Chem 266:21327–21330

Gottschalk S (1990) Biosynthese, Transport und proteolytische Prozessierung der lysosomalen sauren Phosphatase. Dissertation an der Medizinischen Fakultät der Georg-August Universität

Gottschalk S, Waheed A, Schmidt B, Laidler P, von Figura K (1989a) Sequential processing of lysosomal acid phosphatase by a cytoplasmic thiol proteinase and a lysosomal aspartly proteinase. EMBO J 8:3215–3219

Gottschalk S, Waheed A, von Figura K (1989b) Targeting of lysosomal acid phosphatase with altered carbohydrate. Biol Chem Hoppe-Seyler 370:75–80

Hille A, Klumpermann J, Geuze HJ, Peters C, Brodsky FM, von Figura K (1992) Lysosomal acid phosphatase is internalized via clathrin-coated pits. Eur J Cell Biol 59:106–115

Himeno M, Nakamura K, Tanaka Y, Yamada H, Imoto T, Kato K (1991) Mechanisms of a conversion from membrane-associated lysosomal acid phosphatase to content forms. Biochem Biophys Res Commun 180:1483–1489

Johnson KF, Kornfeld S (1992) A His-Leu-Leu sequence near the carboxyl terminus of the cytoplasmic domain of the cation dependent mannose 6-phosphate receptor is necessary for the lysosomal enzyme sorting function. J Biol Chem 267:17110–17115

Kornfeld S (1992) Structure and function of the mannose 6-phosphate/insulin-like growth factor II receptors. Annu Rev Biochem 61:307–330

Kornfeld S, Mellmann I (1989) The biogenesis of lysosomes. Ann Rev Cell Biol 5:483–525

Lehmann LE, Eberle W, Krull S, Prill V, Schmidt B, Sander C, von Figura K, Peters C (1992) The internalization signal in the cytoplasmic tail of lysosomal acid phosphatase consists of the hexapeptide PGYRHV. EMBO J 11:4391–4399

Lemansky P, Gieselmann V, Hasilik A, von Figura K (1985) Synthesis and transport of lysosomal acid phosphatase in normal and I-cell fibroblasts. J Biol Chem 260:9023–9030

Ludwig T, Griffiths G, Hoflack B (1991) Distribution of newly synthesized lysosomal enzymes in the endocytic pathway of normal rat kidney cells. J Cell Biol 115:1561–1572

Parenti G, Willemsen R, Hogeveen AT, Verlema-Mooyman, van Dongen J, Galjaard H (1987) Immunocytochemical localization of lysosomal acid phosphatase in normal and "I-cell" fibroblasts. Eur J Cell Biol 43:121–127

Pearse BMF, Robinson MS (1990) Clathrin, adaptors, and sorting. Ann Rev Cell Biol 6:151–171

Peters C, Braun M, Weber B, Wendland M, Schmidt B, Pohlmann R, Waheed A, von Figura K (1990) Targeting of a lysosomal membrane protein: a tyrosine containing endocytosis signal in the cytoplasmic tail of lysosomal acid phosphatase is necessary and sufficient for targeting to lysosomes. EMBO J 9:3497–3506

Peters C, Geier C, Pohlmann R, Waheed A, von Figura K, Roiko K, Virkkunen P, Vihko P (1989) High degree of homology between primary structures of human lysosomal acid phosphatase and human prostatic acid phosphatase. Biol Chem. Hoppe-Seyler 370:177–181

Pohlmann R, Krentler C, Schmidt B, Schröder W, Lorkowski G, Cully J, Mersmann G, Geier C, Waheed A, Gottschalk S, Grzeschik KH, Hasilik A, von Figura K (1988) Human lysosomal acid phosphatase: cloning, sequencing and chromosomal assignment. EMBO J 7:2343–2350

Prill V, Lehmann L, von Figura K, Peters C (1993) The cytoplasmic tail of lysosomal acid phosphatase contains overlapping but distinct signals for basolateral sorting and rapid internalization in polarized MDCK cells. EMBO J: in press

Schröder S, Ungewickell E (1991) Subunit interaction and function of clathrin-coated vesicles adapters from the Golgi and the plasma membrane. J Biol Chem 266:7910–7918

Sosa M, Schmidt B, von Figura K, Hille-Rehfeld A (1993) In vitro binding of plasma membrane-coated vesicle adapters to the cytoplasmic domain of lysosomal acid phosphatase. J Biol Chem 268:12537–12543

Tanaka Y, Yano S, Furuno K, Ishikawa T, Himeno M, Kato K (1990) Transport of acid phosphatase to lysosomes does not involve passage through the cell surface. Biochem Biophys Res Commun 170:1067–1073

von Figura K, Hasilik A (1986) Lysosomal enzymes and their receptors. Ann Rev Biochem 55:167–193

Waheed A, Gottschalk S, Hille A, Krentler C, Pohlmann R, Braulke T, Hauser H, Geuze H, von Figura K (1988) Human lysosomal acid phosphatase is transported as a transmembrane protein to lysosomes in transfected baby hamster kidney cells. EMBO J 7:2351–2358

Waheed A, van Etten R (1985) Biosynthesis and processing of lysosomal acid phosphatase in cultured human cells. Arch Biochem Biophys 243:274–283

Yokota S, Himeno M, Kato K (1989) Immunocytochemical localization of acid phosphatase in rat liver. Cell Struct Funct 14:163–171

VIP21, Caveolae and Sorting in the *Trans*-Golgi Network of Epithelial Cells

P. Dupree, K. Fiedler, and K. Simons[1]

1 Introduction

The specific characteristics of cellular membranes are a consequence not only of their protein constituents, but also of their lipid composition. It is obvious that the behavior of a membrane protein is dependent upon its lipid environment. The interplay between proteins and lipids during membrane trafficking is, however, an area of research that has been largely neglected. Interest in the glycolipid-dependent sorting of proteins in membranes is an exception, and has provided a paradigm that demonstrates the importance of these interactions. This phenomenon was first proposed during studies of protein and lipid sorting in simple epithelial cells. The apical and basolateral plasma membrane domains of these cells have not only distinct protein compositions, but the various lipid classes vary in their distribution as well (van Meer and Simons 1988; Simons and van Meer 1988). In particular, the apical plasma membrane is enriched in certain glycolipids. This polarity is in part generated in the *trans*-Golgi network (TGN), where the proteins and lipids destined for the individual plasma membrane domains are sorted into distinct vesicle carriers (Griffiths and Simons 1986; Wandinger-Ness et al. 1990).

2 The Hypothesis: Glycolipid Microdomains Are Involved in Protein Sorting in the *Trans*-Golgi Network of Epithelial Cells

It has been proposed by Simons and van Meer (1988) not only that the proteins and lipids are cosorted, but that the processes are also highly interdependent. Glycolipid microdomain formation is the fundamental process on which sorting to the apical plasma membrane domain is based. Owing to their ability to form intermolecular hydrogen bonds (Pascher 1976), these lipids gather into dynamic microdomains within a lipid bilayer. According to the hypothesis, microdomains form the nucleation center around which apically directed proteins cluster. Some proteins may have an affinity for the lipid domains themselves, whereas others would associate indirectly via other proteins. Consequently, the signal for sorting into apical membranes may be either based upon protein-lipid interactions or via less well-defined protein domains that interact with a protein sorter or a previously sorted molecule. In this way, a subset of

[1] Cell Biology Programme, European Molecular Biology Laboratory, Postfach 102209, D-69012 Heidelberg, FRG.

44. Colloquium Mosbach 1993
Glyco- and Cellbiology

lipids and proteins becomes associated in membrane domains (rafts) that bud to form a vesicle (Simons and van Meer 1988; Simons and Wandinger-Ness 1990).

Association of these components into rafts must be both transient and regulated. During passage through the secretory pathway, proteins associate only in the later Golgi complex compartments when they encounter the glycolipids. Importantly, the delivered proteins must be released from the raft at the plasma membrane – the proteins cannot remain clustered as they were in the budding vesicle. Arrival at the plasma membrane should therefore provide a signal for the raft to dissociate.

3 Detergent Insolubility Provides an Insight into Microdomain Formation

The proposal that glycolipids and apically directed proteins interact in this way led us and others to investigate lipid-protein complex formation in the secretory pathway (Kurzchalia et al. 1992; Brown and Rose 1992). It was shown that some apically destined proteins become part of a large insoluble structure in certain detergents. Although there is no direct proof that the glycolipid rafts proposed in the hypothesis constitute this insoluble material, there is now strong evidence to support this contention. The phenomenon of detergent insolubility behaves according to the model of glycolipid-based raft formation in terms of both protein specificity and subcellular localization. The apically destined influenza HA protein becomes insoluble in the detergents TX-100 and CHAPS during its passage through the Golgi complex in MDCK cells, whereas the basolaterally destined VSV-G protein does not (Kurzchalia et al. 1992; Fiedler et al. 1993). Furthermore, GPI-linked proteins, which are sorted apically in MDCK cells, also become incorporated into Triton-X100-insoluble complexes in the Golgi complex (Brown and Rose 1992). By following their oligosaccharide modifications, it is clear that the insolubility of these proteins first occurs in the medial and *trans*-Golgi compartment. Perhaps GPI-linked proteins become incorporated in the medial Golgi, before influenza HA, which seems to become insoluble first in the TGN (Fig. 1). Crucially, both the CHAPS-insoluble material and TX-100-insoluble material contain lipids, and they are enriched in glycosphingolipids relative to other membranes (Brown and Rose 1992; Fiedler et al. 1993). The TX-100-insoluble material appears as membrane fragments when examined in the electron microscope, reinforcing the view that the detergent is unable to solubilize certain membrane domains (Brown and Rose 1992). It has recently been demonstrated using mixed lipid membranes without proteins that the insolubility is partly a consequence of the lipids themselves (Brown 1992). The relationship between the glycolipid microdomains and detergent insolubility, our experimental tool for investigating the phenomenons, is thus now firmly established. Proteins and also lipids will behave differently and be associated with the rafts to different degrees or in different ways. It should be noted that the CHAPS-insoluble complexes have a higher density and contain less lipid than the TX-100-insoluble material (Fiedler et al. 1993). Therefore it must be recognized that solubilization by a detergent does not categorically demonstrate that a protein or a lipid is not part of a raft. This aside,

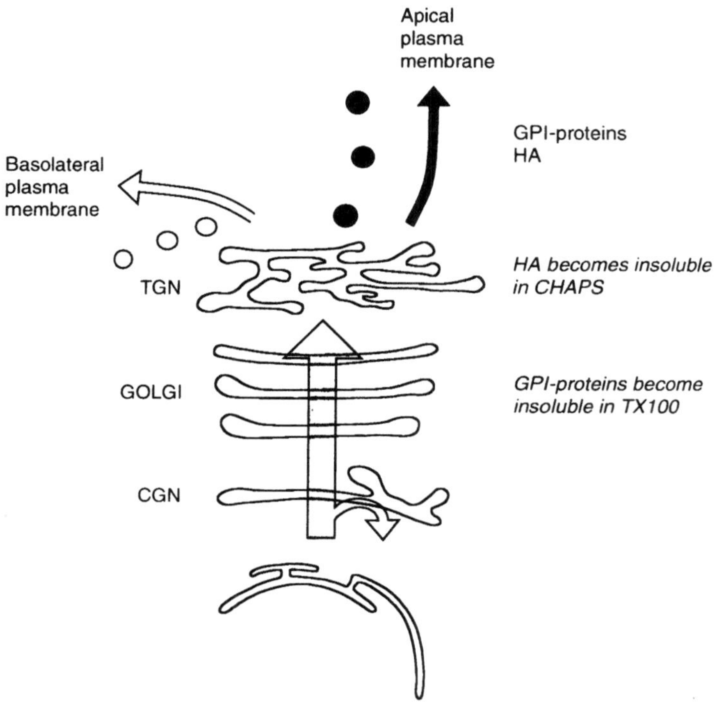

Fig. 1. Schematic depiction of the development of detergent-insoluble complexes during passage through the secretory pathway in a polarized epithelial cell

detergents have become a powerful method to investigate glycolipid microdomain formation.

4 Protein Components of the Glycolipid Rafts in the *Trans*-Golgi Network

The detergent-insoluble domains should contain not only apically transported proteins, but also proteins of interest concerned with protein sorting and vesiculation. Therefore the protein composition of the material has been investigated in some detail. In the first experiments, TGN-derived vesicles from influenza infected cells were solubilized with CHAPS, and the proteins resistant to this treatment were analyzed by two-dimensional gel electrophoresis (Kurzchalia et al. 1992). A subset of the vesicular proteins along with the influenza HA was present, suggesting that these were intimately associated with the glycolipid rafts. The remarkable specificity of the solubilization became clear when the insoluble material from crude membranes was examined; in the low molecular weight range, very few proteins other than those found in the TGN-derived vesicles were present. Although some quantitative differences were evident, the protein composition of the insoluble material was remarkably similar whether CHAPS or TX-114 solubilization procedures were used (Fiedler et al. 1993).

This detergent extraction procedure was used to purify and obtain peptide sequence of one protein present in TGN-derived vesicles from this detergent-insoluble material. A cDNA encoding this protein was cloned and the encoded protein was named VIP21 (Vesicular Integral membrane Protein of 21 kDa; Kurzchalia et al. 1992).

5 VIP21 is Present in Both the *Trans*-Golgi Network and in Caveolae

The localization of a protein component of the detergent-insoluble material is crucial to the hypothesis that this material represents glycolipid rafts of the Golgi complex and TGN-derived vesicles. For this purpose, antisera were raised against two synthetic peptides from the N and C termini of VIP21, and named VIP21-N and VIP21-C respectively (Dupree et al. 1993). In immunofluorescence studies, VIP21-C labeled the Golgi complex of MDCK cells, and colocalized with influenza HA when transport of this protein between the TGN and the plasma membrane was blocked by incubation at 20 °C. Since the protein had been identified originally in TGN-derived vesicles that had been imunoisolated from perforated MDCK cells using antibodies against the HA protein, this was an expected result. Furthermore, it strongly supports the view that the detergent-insoluble material is derived from the TGN and post-Golgi compartments.

As an integral membrane protein transported in vesicles from the TGN to the plasma membrane, VIP21 was, not surprisingly, also present on the plasma membrane. However, immunofluorescent labeling with VIP21-N revealed that the protein was very concentrated in plasma membrane invaginations (Dupree et al. 1993). These were not clathrin-coated, and were very similar to the structures called caveolae (Anderson et al. 1992). Indeed, recent sequence data showing that a chicken protein called caveolin, also localized to caveolae, is homologous to VIP21 confirms this localization (Rothberg et al. 1992; Glenney and Soppet 1992).

6 Does Biogenesis of Caveolae Also Involve Glycolipid Rafts?

The presence of VIP21-caveolin in caveolae raises the question whether caveolae are related to glycolipid rafts of exocytic carrier vesicles. Several lines of evidence suggest that they are. Caveolae probably have a lipid composition different to the bulk of the plasma membrane. Cholera and tetanus toxins bind specifically to gangliosides, and can therefore be used as a marker for membrane domains containing these glycolipids. These toxins have been shown to bind specifically to nonclathrin-coated plasma membrane invaginations (Montesano et al. 1982), suggesting that these domains are enriched in this type of glycolipid. These invaginations are apparently caveolae (R.G. Parton in prep.). Several studies have shown that caveolae are enriched in cholesterol, and that the cholesterol is essential for their integrity (Rothberg et al.

1990b; Chang et al. 1992). However, we await a direct biochemical demonstration that these domains are enriched in glycolipids

GPI-linked proteins are associated both with caveolae and the glycolipid rafts in the Golgi complex. At the plasma membrane, the folate receptor, 5' nucleotidase, Thy-1 and the prion PrP (C) are localized preferentially in caveolae (Rothberg et al. 1990a; Ying et al. 1993). Furthermore, removal of the transmembrane domain of CD4 and addition of a GPI anchor directs this protein to nonclathrin-coated invaginations, which presumably are caveolae (Keller et al. 1992). The presence of VIP21 in the glycolipid rafts of both TGN-derived vesicles and caveolae now functionally links these two membrane subdomains.

Although the localization data do not address the question of endocytosis mediated by these structures themselves, the results do suggest that VIP21-caveolin can be internalized. The VIP21-C antibody labels the TGN even after 10 h of influenza viral infection when protein synthesis has been shut off (Dupree et al. 1993). Since the protein continually leaves the Golgi together with HA protein in TGN-derived vesicles during this time, it seems probable that the protein might recycle from the plasma membrane back to the TGN. VIP21-caveolin could be internalized via coated pits, but localization studies showed that it is virtually excluded from these domains (Dupree et al. 1993). There is considerable evidence that an endocytic pathway exists with different characteristics to the clathrin-mediated pathway, having different kinetics and not being affected by certain inhibitors (Van Deurs et al. 1989). Whether this process involves caveolae is unclear, but a few observations suggest that it may do. Indeed, cholera and tetanus toxins (Montesano et al. 1982; Tran et al. 1987) and GPI-linked proteins (Keller et al. 1992) can apparently be internalized by this clathrin-independent pathway.

7 A General Model of Glycolipid Microdomain-Based Sorting

We must now consider glycolipid raft-based sorting as a general phenomenon that mediates segregation of proteins into subdomains of a membrane. The similarity between the processes of protein and lipid sorting that are occurring in the TGN and in the biogenesis of caveolae suggests that the phenomenon is more widespread.

VIP21 is present in both apical and basolateral vesicles of MDCK cells (Wandinger-Ness et al. 1990; Kurzchalia et al. 1992). Furthermore, there are caveolae containing VIP21 in the basolateral membrane (Dupree et al. 1993). Glycolipids are routed not only apically but also basolaterally, although to a lesser extent (Simons and van Meer 1988). VIP21 therefore may be a specific marker of glycolipid rafts in both TGN-derived vesicles and caveolae. The pressing question is the identity of proteins that determine the specific association with the rafts present in caveolae or apical vesicles which distinguishes these processes. We can speculate that there are specific sorting proteins that have an affinity for certain lipids and are regulated by a compartment-specific pH or ionic composition. In this way, apically destined proteins are induced to associate with apically destined glycolipid microdomains and VIP21 in the TGN. Association with the glycolipids in the TGN that move basolaterally may be prevented by the absence of sorting proteins, or alternatively these sorting proteins may be posttranslationally modified to become inactive. Upon arrival at the plasma membran, the cargo is released, but caveolae-specific proteins are induced to associate with the glycolipid microdomains. These may include proteins excluded from the apically destined rafts at the TGN. One particularly clear example is provided by the β-adrenergic receptor. This protein is directed basolaterally in MDCK cells (Simmons et al. 1984), and therefore does not associate with the apically directed rafts in the TGN. In contrast, after modulation with an antibody, it can accumulate very specifically in caveolae with VIP21 (Dupree et al. 1993). This demonstrates clearly the principle that the association of a protein with these rafts is regulated.

If glycolipid raft-based sorting is not limited to the TGN, can we find yet further examples of this phenomenon? Some observations on T-cell activation suggest that this may be the case, and point to intriguing areas that should be pursued in the future. It is possible to activate T-cells with antibodies against GPI-linked proteins (Robinson 1991), an observation without an obvious explanation, since the src signaling molecules and GPI-linked proteins are without transmembrane domains for a direct interaction. Furthermore, src family kinases can be co-immunoprecipitated with several GPI-linked proteins after solubilisation with nonionic detergents (Stefanova et al. 1991; Cinek and Horejsi 1992). This might be explained by an indirect interaction via a transmembrane coupling protein. However, the detergent-insoluble material contains glycolipids, and is remarkably similar in size to the detergent-insoluble material from ephitelial cells (Cinek and Horejsi 1992). It thus seems highly likely that the clustering of GPI-linked proteins and src is mediated by a glycolipid raft-like process (Dupree et al. 1993). The relationship between these plasma membrane domains and caveloae is not clear, but there is no reason a priori to expect that these rafts have a caveolar morphology. This morphological appearance of the domains may be regulated by specific proteins. It will be interesting to see whether this detergent-insoluble material also contains VIP21. On this point, VIP21-caveolin was identified as a protein phosphorylated by src in v-src transformed cells (Glenney 1989), also suggesting that src may be held in a close association with VIP21 in glycolipid-enriched microdomains. These observations have convinced us that glycolipid-enriched domains will soon be shown to be an important feature of several membrane processes.

Acknowledgments. We thank Rob Parton for useful discussions. The work was supported by The Royal Society (P.D.), Boehringer Ingelheim Fonds (K.F.), and Sonderforschungsbereich 352 of the Deutscheforschungsgemeinschaft (K.S.).

References

Anderson RGW, Kamen BA, Rothberg KG, Lacey SW (1992) Potocytosis: sequestration and transport of small molecules by caveolae. Science 255:410–411

Brown DA (1992) Interactions between GPI-anchored proteins and membrane lipids. Trends Cell Biol 2:338–343

Brown D, Rose J (1992) Sorting of GPI-anchored proteins to glycolipid-enriched membrane subdomains during transport to the apical cell surface. Cell 68:533–544

Chang W-J, Rothberg KG, Kamen BA, Anderson RGW (1992) Lowering the cholesterol content of MA104 cells inhibits receptor-mediated transport of folate. J Cell Biol 118:63–69

Cinek T, Horejsi V (1992) The nature of large noncovalent complexes containing Glycosyl-phosphatidyl inositol-anchored membrane glycoproteins and protein tyrosine kinases. J Immunol 149:2262–2270

Dupree P, Parton RG, Raposo G, Kurzchalia TV, Simons K (1993) Caveolae and sorting in the trans-Golgi network of epithelial cells. EMBO J 12:1597–1604

Fiedler K, Kobayashi T, Kurzchalia TV, Simons K (1993) Glycosphingolipid-enriched detergent-insoluble complexes in protein sorting in ephithelial cells. Biochemistry 32:6365–6373

Glenney JR (1989) Tyrosine phosphorylation of a 22-kDa protein is correlated with transformation by rous sarcoma virus. J Biol Chem 264:20163–20166

Glenney JR, Soppet D (1992) Sequence and expression of caveolin, a protein component of caveolae plasma membrane domains phosphorylated on tyrosine in Rous sarcoma virus-transformed fibroblasts. Proc Nat Acad Sci USA 89:10517–10521

Griffiths G, Simons K (1986) The *trans*-Golgi network: sorting at the exit site of the Golgi complex. Science 234:438–443

Keller GA, Siegel MW, Caras IW (1992) Endocytosis of glycophospholipid-anchored and transmembrane forms of CD4 by different endocytic pathways. EMBO J 11:863–874

Kurzchalia TV, Dupree P, Parton RG, Kellner R, Virta H, Lehnert M, Simons K (1992) VIP21, A 21-kD membrane protein is an integral component of *trans*-Golgi network-derived transport vesicles. J Cell Biol 118:1003–1014

Montesano R, Roth J, Robert A, Orci L (1982) Non-coated membrane invaginations are involved in binding and internalisation of cholera and tetanus toxins. Nature 296:651–653

Pascher I (1976) Molecular arrangements in sphingolipids. Conformation and hydrogen bonding of ceramide and their implication on membrane stability and permeability. Biochim Biophys Acta 455:433–451

Robinson PJ (1991) Phosphatidylinositol membrane anchors and T-cell activation. Immunology today 12:35–41

Rothberg KG, Ying Y, Kolhouse JF, Kamen BA, R.G.W.A (1990a) The glycophospholipid-linked folate receptor internalizes folate without entering the clathrin-coated pit endocytic pathway. J Cell Biol 110:637–649

Rothberg KG, Ying Y-S, Kamen BA, Anderson RGW (1990b) Cholesterol controls the clustering of the glycophospholipid-anchored membrane receptor for 5-methyltetrahydrofolate. J Cell Biol 111:2931–2938

Rothberg KG, Heuser JE, Donzell WC, Ying Y-S, Glenney JR, Anderson RGW (1992) Caveolin, a protein component of caveolae membrane coats. Cell 68:673–682

Simmons NL, Brown CDA, Rugg EL (1984) The action of epinephine on Madin-Darby canine kidney cells. Fed Proc 43:2225–2229

Simons K, van Meer G (1988) Lipid sorting in epithelial cells. J Biochem 27:6197–6202

Simons K, Wandinger-Ness A (1990) Polarized sorting in ephitelia. Cell 62:207–210

Stefanova I, Horejsi V, Ansotegui IJ, Knapp W, Stockinger H (1991) GPI-anchored cell-surface molecules complexed to protein tyrosine kinases. Science 254:1016–1019

Tran D, Carpentier J-L, Sawano F, Gorden P, Orci L (1987) Ligands internalized through coated or non-coated invaginations follow a common intracellular pathway. Proc Natl Acad Sci USA 84:7957–7961

Van Deurs B, Petersen OW, Olsnes S, Sandvig K (1989) The ways of endocytosis. Int Rev Cytol 117:131–178

van Meer G, Simons K (1988) Lipid polarity and sorting in ephitelial cells. J Cell Biochem 36:51–58.

Wandinger-Ness A, Bennett MK, Antony C, Simons K (1990) Distinct transport vesicles mediate the delivery of plasma membrane proteins to the apical and basolateral domains of MDCK cells. J Cell Biol 111:987–1000

Ying Y-S, Anderson RGW, Rothberg KG (1993) Each caveola contains multiple GPI-linked proteins. Cold Spring Harbor Symposia on Quantitative Biology. Cold Spring Habor Laboratory Press, Vol. 57, 593–604

Biogenesis of Neurosecretory Vesicles

W. B. Huttner[1], F. A. Barr[2], R. Bauerfeind[1], O. Bräunling[1], E. Chanat[3], A. Leyte[4], M. Ohashi[1], A. Régnier-Vigouroux[5], T. Flatmark[6], H.-H. Gerdes[1], P. Rosa[7], and S. A. Tooze[8]

1 Introduction

Neurosecretory vesicles are defined as the vesicles which mediate the usually calcium-dependent, regulated release of signaling molecules in the nervous system. At least two types of neurosecretory vesicles can be distinguished by their structure and content. The first type are the synaptic vesicles of neurons, which mediate the storage and release of classical neurotransmitters (e.g., acetylcholine, glutamate, GABA, and glycine) but lack secretory proteins. Synaptic vesicles have a counterpart in certain endocrine cells, the synaptic-like microvesicles (SLMVs). The second type are the large dense core vesicles of neurons, which mediate the storage and release of neuropeptides. These vesicles are the neuronal equivalent of the secretory granules found in cells capable of regulated protein secretion, notably endocrine cells. (Since large dense cote vesicles are essentially similar, if not identical, to endocrine secretory granules, we shall also use the term "secretory granules" when refering to large dense core vesicles of neurons.) This chapter summarizes recent data obtained in our laboratory concerning the biogenesis of secretory granules from the *trans*-Golgi network (TGN) and of SLMVs from early endosomes in the neuroendocrine cell line PC12.

[1] Institute for Neurobiology, University of Heidelberg, Im Neuenheimer Feld 364, D-69120 Heidelberg, FRG.
[2] Present address: National Institute for Medical Research, The Ridgeway, Mill Hill, London NW7 1AA, UK.
[3] Present address: Laboratoire de Biologie Cellulaire et Moléculaire, INRA, F-78352 Jouy-en-Josas Cédex, France.
[4] Present address: The Netherland Cancer Institute, Plesmanlaan 121, 1066 CX Amsterdam, The Netherlands.
[5] Present address: Centre d'Immunologie INSERM-CNRS de Marseille-Luminy, F-13288 Marseille Cédex 9, France.
[6] Present address: Department of Biochemistry, University of Bergen, N-5009 Bergen, Norway
[7] CNR Center of Cytopharmacology, Department of Pharmacology, School of Medicine, Via Vanvitelli 32, I-20129 Milan, Italy.
[8] European Molecular Biology Laboratory, Meyerhofstr. 1, D-69117 Heidelberg, FRG.

44. Colloquium Mosbach 1993
Glyco- and Cellbiology

2 Biogenesis of Secretory Granules

2.1 Regulation by Heterotrimeric G Proteins

Using a cell-free system derived from PC12 cells, which reconstitutes the formation of secretory granules from the TGN (formation being defined as the physical detachment of newly budded secretory granules from the TGN) (Tooze and Huttner 1990), we had previously observed that this process is inhibited by nonhydrolyzable analogues of GTP, suggesting an involvement of GTP-binding proteins (Tooze et al. 1990). Work carried out during the past 2 years has shown that one class of GTP-binding protein involved in secretory granule formation are the heterotrimeric G proteins. The first indication for this came from experiments which examined the effects of aluminum fluoride, an activator of heterotrimeric but not small ras-like GTP-binding proteins, and of purified heterotrimeric G protein $\beta\gamma$ subunits in the cell-free system. Aluminum fluoride was found to inhibit secretory granule formation to the same extent as the nonhydrolyzable GTP analogue GTPγS (Barr et al. 1991). Conversely, addition of purified $\beta\gamma$ subunits stimulated the formation of secretory granules (Barr et al. 1991). Further work showed that multiple heterotrimeric G proteins of the Gi/Go and Gs class are associated with the TGN (Leyte et al. 1992). Activation of Gi/Go by mastoparan, a peptide which mimicks the effect of an activated receptor on Gi/Go by promoting guanine nucleotide exchange on these G proteins, results in the inhibition of secretory granule formation from the TGN in the cell-free system. Pretreatment of cells with pertussis toxin, which renders Gi/Go unable to interact with receptors, not only blocks this effect of mastoparan but by itself results in an increase in cell-free secretory vesicle formation (Leyte et al. 1992). These findings suggest that guanine nucleotide exchange factors operate on Gi/Go in the cell-free system, hinting at the existence of receptors which act in such a manner. Activation of Gs by cholera toxin, on the other hand, stimulates cell-free secretory vesicle formation. Thus, Gi/Go and Gs exert opposing regulatory effects on secretory vesicle formation from the TGN (Leyte et al. 1992).

2.2 Coat Proteins

The effector systems through which the various heterotrimeric G proteins exert their stimulatory and inhibitory effects on secretory granule formation are unknown. However, two lines of evidence suggest that coat proteins, implicated in vesicle formation at other sites in the endomembrane system, are also involved in the formation of secretory granules from the TGN and may be targets for regulation by heterotrimeric G proteins. The first line of evidence comes from a study on the effect of brefeldin A on secretory granule formation. Brefeldin A, a fungal metabolite that inhibits vesicular transport along the secretory pathway (Klausner et al. 1992), prevents the membrane binding of components of the nonclathrin and clathrin coat, which are thought to be required for the budding of vesicles that mediate biosynthetic protein traffic (Rothman and Orci 1992). Brefeldin A has been found to inhibit the formation of secretory granules in vivo (Miller et al. 1992; Rosa et al. 1992) and in the cell-free system (Rosa et al. 1992). If we assume a common mechanism of action for brefeldin A irrespective of the donor compartment affected by the drug, the inhibition of secretory granule formation from the TGN by brefeldin A suggests that coat protein binding to the TGN is prevented by the drug, and that this is the underlying cause for the inhibition of secretory granule formation. It will be important to determine whether the clathrin coat of immature secretory granules or other coat proteins are the targets for brefeldin A that are relevant in this context.

The second line of evidence comes from a study on the effect of ARF (ADP-ribosylation factor), a small GTP-binding protein, on secretory granule formation. We have found that a synthetic, N-terminally myristylated peptide corresponding to the N-terminal domain of ARF-1 stimulates the formation of secretory granules in the cell-free system, whereas the corresponding nonmyristylated peptide, or peptides corresponding to the N-terminal domain of ARF-4, do not (Barr and Huttner in prep.). Thus, ARF-1 or an ARF-1-like protein may be involved in secretory granule formation.

If one extrapolates from the results obtained by other investigators using Golgi membranes, it may be that the effects of brefeldin A and of the ARF-1 peptide on cell-free secretory granule formation from the TGN are connected. Brefeldin A inhibits an activity associated with Golgi membranes that catalyzes the guanine nucleotide exchange on ARF (Donaldson et al. 1992b; Helms and Rothman 1992), which is a prerequisite for its binding to membranes (Donaldson et al. 1992a). This, in turn, is required for the membrane binding of other coat components (Donaldson et al. 1992a). Thus, by analogy with the formation of vesicles mediating transport through the Golgi complex, the sequence of events in secretory granule formation from the TGN may be: TGN-associated catalysis of guanine nucleotide exchange on ARF → binding of ARF to TGN membranes → binding of coat proteins to the TGN → budding of immature secretory granules, with brefeldin A inhibiting the first event.

2.3 Sorting Signals

We have previously identified a highly conserved domain common to two widespread regulated secretory proteins of neurons and endocrine cells, chromogranin A and chromogranin B (secretogranin I; Huttner et al. 1991), which consists of a disulfide-bonded 20-amino-acid-long loop structure encoded by a separate exon (Benedum et al. 1987; Pohl et al. 1990). Exploiting the observations by other investigators that disulfide bond formation can be inhibited in living cells by the addition of thiol reducing agents (Alberini et al. 1990; Braakman et al. 1992), the role of this intramolecular disulfide bond in the sorting of chromogranin B to secretory granules of PC12 cells has been studied (Chanat et al. 1993). Reduction of the disulfide bond resulted in the selective missorting of chromogranin B to constitutive secretory vesicles, whereas the sorting of the related regulated secretory protein secretogranin II, which lacks disulfide bonds, was unaffected. The reduced chromogranin B still exhibited the low pH/calcium-induced aggregation implicated in its sorting (Chanat et al. 1993; Chanat and Huttner 1991). These data therefore suggest that aggregation, although in principle capable of causing the segregation of regulated from constitutive secretory proteins within the lumen of the TGN (Chanat et al. 1991; Tooze et al. 1993), is not alone sufficient to allow the sorting process to go to completion, i.e., to generate secretory granules from the TGN. Rather, in the case of chromogranin B, the sorting process can only be completed if the structural information associated with the disulfide bond is preserved. It is tempting to speculate that this structural information may be required for the interaction of soluble chromogranin B with receptors in the TGN membrane involved in sorting.

2.4 Secretory Granule Maturation

Immature secretory granules have previously been shown to be a short-lived intermediate in the biogenesis of mature secretory granules (Tooze et al. 1991; Tooze et al. 1993). The characterization of the fate of immature secretory granules in PC12 cells has revealed that there are at least two maturation routes. Immature secretory granules may mature (1) into larger secretory granules by self-fusion (Tooze et al. 1991), or (2) into smaller secretory granules by condensation of the contents (Bauerfeind et al. 1993), accompanied in both cases by the removal of the excess membrane. It will be important to determine the fate of the excess membrane removed from immature secretory granules during their maturation. This membrane may be used (1) for the recycling of sorting receptors to the TGN, (2) for the segregation of lysosomal proteins and residual constitutive secretory proteins from regulated secretory proteins, or (3) for the removal of certain products of proteolytic processing of regulated secretory proteins (Bauerfeind and Huttner 1993). Whatever the fate of the membrane may be, the immature secretory granule may well turn out to be a post-TGN sorting compartment on the biosynthetic secretory pathway.

3 Biogenesis of SLMVs

3.1 SLMVs Originate from Early Endosomes

We have previously shown that in PC12 cells, newly synthesized synaptophysin, a major synaptic vesicle membrane protein, travels in constitutive secretory vesicles from the TGN to the plasma membrane, cycles several times between the plasma membrane and early endosomes, and is then packaged into SLMVs (Régnier-Vigouroux et al. 1991). This suggested that the de novo formation of SLMVs occurs from early endosomes, although their formation directly from the plasma membrane was not strictly excluded. Recent experiments extending this work show that the fluid phase marker horseradish peroxidase, pre-internalized into early endosomes, is in part chased to SLMVs, thus providing evidence for the formation of SLMVs directly from early endosomes (Bauerfeind et al. 1993).

3.2 SLMVs and Early Endosomes Take Up and Store Classical Neurotransmitters

It has been suggested, but not unequivocally shown, that SLMVs store classical neurotransmitters (Reetz 1991). Using metabolic labeling of PC12 cells with radioactive choline followed by subcellular fractionation and analysis of acteylcholine, we have recently shown that SLMVs do indeed contain biosynthetic acetylcholine (Bauerfeind et al. 1993). Interestingly, a substantial portion of the biosynthetic acetylcholine is found in early endosomes, and both early endosomes and SLMVs contain an ATP-dependent acetylcholine uptake system (Bauerfeind et al. 1993). These findings are consistent with the early endosomal origin of SLMVs.

3.3 SLMVs and the Storage of Biogenic Amines

Because PC12 cells are capable of neurotransmitter uptake and storage in SLMVs, and also express the uptake system for biogenic amines such as catecholamines, one can use these cells to address the question whether or not SLMVs also contain catecholamines. Our experiments indicate that, at least for undifferentiated PC12 cells, this is not the case (Bauerfeind et al. 1993). In these cells, catecholamine storage is confined to mature secretory granules, with no detectable storage in SLMVs. Consistent with this, the vesicular amine transporter is present in the membrane of mature secretory granules. Interestingly, the vesicular amine transporter is also detected in early endosomes, but not in SLMVs (Bauerfeind et al. 1993). If our interpretation that this reflects the recycling of the amine transporter after exocytosis of secretory granules to the TGN is correct, then this implies that the early endosomes of these PC12 cells differentially sort two neurotransmitter uptake systems, the acetylcholine transporter to SLMVs and the amine transporter to vesicles destined to the TGN.

The finding that SLMVs of undifferentiated PC12 cells lack catecholamines has implications about the nature of the so-called small dense core vesicles (SDCVs) of sympathetic neurons, which lack secretory proteins, contain catecholamines, and because of their small size have been considered to be related to the synaptic vesicles. The latter notion is supported by our observation (Bauerfeind and Huttner in prep.) that SDCVs contain synaptophysin. If SDCVs should indeed prove to be catecholamine-storing synaptic vesicles, this would imply that the vesicular amine transporter expressed in neurons is sorted differently than the one expressed in undifferentiated PC12 cells. This possibility offers an interesting perspecitve to identify sorting signals underlying the traffic of membrane proteins to synaptic vesicles.

4 Conclusions

(1) The formation of secretory granules from the TGN is regulated by multiple heterotrimeric G proteins in a stimulatory (Gs) and inhibitory (Gi/o) fashion. (2) Effector systems of the TGN-associated heterotrimeric G proteins may include ARF and other coat proteins. (3) Receptor system stimulating guanine nucleotide exchange on the

TGN-associated heterotrimeric G proteins may recognize specific structures in regulated secretory proteins such as the conserved disulfide-bonded loop of chromogranin B which appears to contain sorting information. (4) The membrane retrieval from immature secretory granules occurring during secretory granule maturation may provide a mechanism for recycling of sorting receptors to the TGN. (5) Membrane proteins destined to SLMVs travel from the TGN via the constitutive secretory pathway to the plasma membrane and from there to early endosomes, the compartment from which the de novo formation of SLMVs occurs. (6) SLMVs contain classical neurotransmitters (acetylcholine) but, at least in undifferentiated PC12 cells, lack catecholamines, despite the presence of both the acetylcholine transporter and vesicular monoamine transporter in early endosomes, implying differential sorting of two distinct neurotransmitter uptake systems in this compartment.

Acknowledgments. WBH was the recipient of a grant from the Deutsche Forschungsgemeinschaft (SFB 317).

References

Alberini CM, Bet P, Milstein C, Sitia R (1990) Secretion of immunoglobulin M assembly intermediates in the presence of reducing agents. Nature 347:485–487

Barr FA, Leyte A, Mollner S, Pfeuffer T, Tooze SA, Huttner WB (1991) Trimeric G-proteins of the *trans*-Golgi network are involved in the formation of constitutive secretory vesicles and immature secretory granules. FEBS Lett 294:239–243

Bauerfeind R, Huttner WB (1993) Biogenesis of constitutive secretory vesicles, secretory granules and synaptic vesicles. Curr Opinion Cell Biol 5 (in press)

Bauerfeind R, Régnier-Vigouroux A, Flatmark T, Huttner WB (1993) Selective storage of acetylcholine, but not catecholamines, in neuroendocrine synaptic-like microvesicles of early endosomal origin. Neuron (in press)

Benedum UM, Lamouroux A, Konecki DS, Rosa P, Hille A, Baeuerle PA, Frank R, Lottspeich F, Mallet J, Huttner WB (1987) The primary structure of human secretogranin I (chromogranin B): comparison with chromogranin A reveals homologous terminal domains and a large intervening variable region. EMBO J 6:1203–1211

Braakman I, Helenius J, Helenius A (1992) Manipulating disulfide bond formation and protein folding in the endoplasmic reticulum. EMBO J 11:1717–1722

Chanat E, Huttner WB (1991) Milieu-induced, selective aggregation of regulated secretory proteins in the *trans*-Golgi network. J Cell Biol 115:1505–1519

Chanat E, Pimplikar SW, Stinchcombe JC, Huttner WB (1991) What the granins tell us about the formation of secretory granules in neuroendocrine cells. Cell Biophysics 19:85–91

Chanat E, Weiß U, Huttner WB, Tooze SA (1993) Reduction of the disulfide bond of chromogranin B (secretogranin I) in the *trans*-Golgi network causes its missorting to the constitutive secretory pathway. EMBO J 12:2159–2168

Donaldson JG, Cassel D, Kahn RA, Klausner RD (1992a) ADP-ribosylation factor, a small GTP-binding protein, is required for binding of the coatomer protein β-COP to Golgi membranes. Proc Natl Acad Sci USA 89:6408–6412

Donaldson JG, Finazzi D, Klausner RD (1992b) Brefeldin A inhibits Golgi membrane-catalysed exchange of guanine nucleotide onto ARF protein. Nature 360:350–352

Helms JB, Rothman JE (1992) Inhibition by brefeldin A of a Golgi membrane enzyme that catalyses exchange of guanine nucleotide bound to ARF. Nature 360:352–354

Huttner WB, Gerdes H-H, Rosa P (1991) Chromogranins/secretogranins – widespread constituents of the secretory granule matrix in endocrine cells and neurons. In: Markers for neural and endocrine cells. Molecular and cell biology, diagnostic applications. M. Gratzl and K. Langley, eds. (Weinheim: VCH), pp 93–131

Klausner RD, Donaldson JG, Lippincott-Schwartz J (1992) Brefeldin A: insights into the control of membrane traffic and organelle structure. J Cell Biol 116:1071–1080

Leyte A, Barr FA, Kehlenbach RH, Huttner WB (1992) Multiple trimeric G-proteins on the *trans*-Golgi network exert stimulatory and inhibitory effects on secretory vesicle formation. EMBO J 11:4795–4804

Miller SG, Carnell L, Moore HPH (1992) Post-Golgi membrane traffic: brefeldin A inhibits export from distal Golgi compartment to the cell surface but not recycling. J Cell Biol 118:267–283

Pohl TM, Phillips E, Song K, Gerdes H-H, Huttner WB, Rüther U (1990) The organisation of the mouse chromogranin B (secretogranin I) gene. FEBS Lett 262:219–224

Reetz A, Solimena M, Matteoli M, Folli F, Takei K, De Camilli P (1991) GABA and pancreatic β-cells: colocalization of glutamic acid decarboxylase (GAD) and GABA with synaptic-like microvesicles suggests their role in GABA storage and secretion. EMBO J 10:1275–1284

Régnier-Vigouroux A, Tooze SA, Huttner WB (1991) Newly synthesized synaptophysin is transported to synaptic-like microvesicles via constitutive secretory vesicles and the plasma membrane. EMBO J 10:3589–3601

Rosa P, Barr FA, Stinchcombe JC, Binacchi C, Huttner WB (1992) Brefeldin A inhibits the formation of constitutive secretory vesicles and immature secretory granules from the trans-Golgi network. Eur J Cell Biol 59:265–274

Rothman JE, Orci L (1992) Molecular dissection of the secretory pathway. Nature 355:409–415

Tooze S, Flatmark T, Tooze J, Huttner WB (1991) Characterization of the immature secretory granule, an intermediate in granule biogenesis. J Cell Biol 115:1491–1503

Tooze SA, Chanat E, Tooze J, Huttner WB (1993) Secretory granule formation. In: Peng Loh Y (ed) Mechanisms of intracellular trafficking and processing of proproteins. CRC Press, Boca Raton, pp 157–177

Tooze SA, Huttner WB (1990) Cell-free sorting to the regulated and constitutive secretory pathways. Cell 60:837–847

Tooze SA, Weiss U, Huttner WB (1990) Requirement for GTP hydrolysis in the formation of secretory vesicles. Nature 347:207–208

Lipid Transport from the Golgi Complex to the Plasma Membrane of Epithelial Cells

G. van Meer, M. Thielemans, I. L. van Genderen, A. L. B. van Helvoort, P. van der Bijl, and K. N. J. Burger[1]

1 Introduction

Eukaryotic cells are enveloped by a plasma membrane. The cytoplasm is filled with numerous intracellular membranes that sometimes enclose also other membraneous systems. It is well recognized that the various organelles exert different functions and possess unique protein and lipid compositions. The intracellular lipid heterogeneity is very stable and strikingly similar among the different cell types. A fundamental question is how it is generated by the dynamic interplay between local synthesis, modification, and degradation on the one hand, and the various modes of lipid traffic on the other.

A special case of two membranes with different composition is found in the two domains of the plasma membrane of epithelial cells. These two domains display unique protein and lipid compositions (Simons and van Meer 1988). The general feature of the lipid polarity appears to be an enrichment of glycosphingolipids on the apical, and of the phospholipid phosphatidylcholine (PC) on the basolateral surface. The difference is maintained by the tight junctions, the zone of cell-cell contacts that encircles the apex of each epithelial cell. This structure acts as a barrier to lipid diffusion in the outer leaflet of the plasma membrane (van Meer and Simons 1986), and the differences in lipid composition must therefore reside in the outer leaflets of the apical and basolateral plasma membrane domain. While the compositional differences appear to be generated by sorting of newly synthesized lipids before they reach the cell surface, lipid sorting along the transcytotic route would also seem necessary to prevent lipid intermixing by that pathway.

The present chapter discusses sphingolipid synthesis in the Golgi, sphingolipid traffic from their site of synthesis to the plasma membrane, and the localization of the sorting events. It evaluates our working hypothesis on the microdomain mechanism of lipid sorting with primary emphasis on the appearance of lipids at the cell surface. Cellular lipid traffic will undoubtedly be more complicated than any scheme provided here. However, the complications may turn out to be variations on a common theme.

[1] Department of Cell Biology, Medical School AZU H02.314, Heidelberglaan 100, 3584 CX Utrecht, The Netherlands.

44. Colloquium Mosbach 1993
Glyco- and Cellbiology

2 Sphingolipid Synthesis in the Golgi Complex

Any single cell expresses a specific set of glycosphingolipids, usually one or two series derived from ceramide by the stepwise addition of monosaccharides. Apart from the addition of the first glucose to the ceramide backbone (Coste et al. 1986; Futerman and Pagano 1991; Trinchera et al. 1991; Jeckel et al. 1992) and the first galactose to the ceramide backbone (K.N.J. Burger, P. van der Bijl, and G. van Meer in prep.), assembly of the glycosphingolipids occurs on the lumenal surface of subsequent cisternae of the Golgi complex. The distribution of the various enzymes over the Golgi is listed in Table 1. Small fractions of SM synthase (Futerman et al. 1990; Kallen et al. 1993; van Helvoort et al. 1993) and protein glycosyltransferase activities (Shur 1989) have been found at the cell surface. The implications for sphingomyelin (SM) and glycosphingolipid biosynthesis are unclear.

SM is synthesized by the energy-independent transfer of phosphorylcholine from a PC molecule onto ceramide. This occurs on the lumenal surface of the Golgi (Futerman et al. 1990; Helms et al. 1990; Jeckel et al. 1990; Karrenbauer et al. 1990; Jeckel et al. 1992), and apparently on the *cis*-aspect of that organelle (Futerman et al. 1990; Jeckel et al. 1990; Jeckel et al. 1992) (Table 1). However, some uncertainties concerning the exact location of the SM synthase persist. If indeed it was located in the *cis*-Golgi it would have been expected to be redistributed to the ER in the presence of brefeldin A (BFA; Table 1). In a direct experiment, this was observed for glucosyltransferase (Strous et al. 1993) but not for SM synthase (in HepG2 cells; van

Table 1. Distribution of SM synthase and glycosyltransferases over the Golgi stack[a]

Enzyme	Precursor	Product	Cell fractionation	Still works with BFA	Still works with Mitotic cells
SM synthase	Ceramide	SM	*cis*[2,3]	+[4,5]	+[6]
GlcT	Ceramide	GlcCer	*cis*[7,8]	+[9–11]	+[6]
GalT-2	GlcCer	LacCer		+[9–11]	+[6]
SAT-1	LacCer	G_{M3}	*cis*[12–14]	+[9–11]	
SAT-2	G_{M3}	G_{D3}	*cis*/medial/*trans*[12,13]	+[9–11]	
GlcNAcT	LacCer	Lc_3	*cis*[15]	+[10]	
GalNAcT	LacCer/G_{M3}/G_{D3}	G_{A2}/G_{M2}/G_{D2}	medial[13]	–[9,11]	–[6]
GalT-6	Lac Cer	Gb_3		–[10]	
GalT-4	Lc_3	nLc_4	*trans*[15]	–[10]	
GalT-3	G_{A2}	G_{A1}	*trans*[13]		
SAT-4/-5	G_{A1}/G_{M1b}	G_{M1b}/G_{D1c}	*trans*[12–14]		

[a] For enzyme nomenclature see (van Meer and Burger 1992); "Still works with": with BFA the Golgi stack, but not the *trans*-Golgi network, is thought to fuse with the ER (Klausner et al. 1992). Newly synthesized ceramide can be glycosylated by the sequential action of the glycosyltransferases that have returned to the ER. GalNAcT is apparently located beyond the cisternae that relocate. The same reasoning may apply to mitotic cells (Thyberg and Moskalewski, 1992); [2](Futerman et al. 1990); [3](Jeckel et al. 1990); [4](Brüning et al. 1992); [5](Hatch and Vance 1992); [6](Collins and Warren 1992); [7](Futerman and Pagano 1991); [8](Jeckel et al. 1992); [9](van Echten et al. 1990); [10](Sherwood and Holmes 1992); [11](Young et al. 1990); [12](Iber et al. 1992); [13](Trinchera et al. 1990); [14](Trinchera and Ghidoni 1989); [15](Holmes 1989).

Meer and van't Hof 1993). Redistribution to the ER would result in an increase in SM synthesis in the presence of BFA. This was observed in some cells types (Brüning et al. 1992; Hatch and Vance 1992) but not in others (van Echten et al. 1990).

3 Sphingolipid Transport to the Cell Surface

After synthesis in the Golgi, glycosphingolipids and SM are transported to the plasma membrane with half-times of 20–30 min, a process which shows a temperature dependence consistent with vesicular transport (Fig. 1) (Miller-Podraza and Fishman 1982; Miller-Prodraza and Fishman 1984; Lipsky and Pagano 1985; van Meer et al. 1987; Helms et al. 1990; Karrenbauer et al. 1990; van't Hof and van Meer 1990; Young et al. 1992; van Meer and van't Hof 1993). The appearance of SM and glycosphingolipids on the cell surface seemed completely inhibited in mitotic cells, supporting a vesicular transport process for both lipids, as vesicular protein transport is reversibly inhibited during mitosis (Kobayashi and Pagano 1989). Another argument in favor of vesicular transport is that transport of both lipids to the apical plasma membrane domain of epithelial cells was inhibited to the same extent (50%) by microtubule depolymerization (van Meer and van't Hof 1993), just like the transport of membrane proteins and secretory proteins. Transport of glucosylceramide (GlcCer) to the

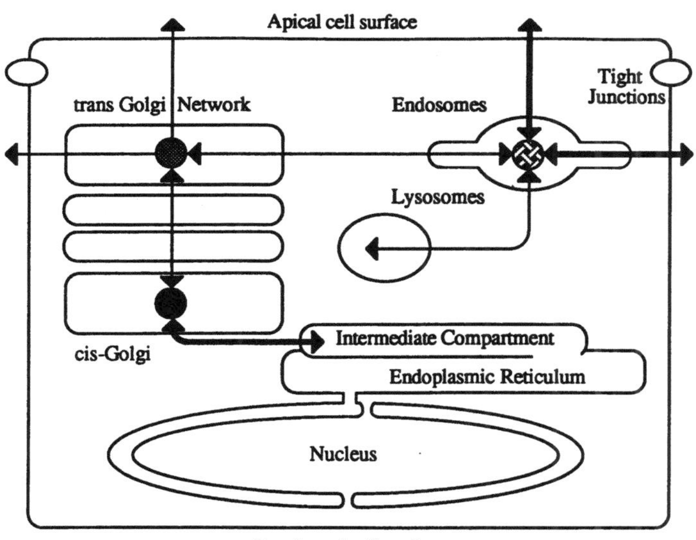

Fig. 1. Vesicular transport pathways in the mammalian cell. The organelles of the vacuolar system are connected by vesicular traffic as indicated by the *arrows*. Excluded from this traffic are the mitochondria and peroxisomes. The *filled circles* represent sites where membrane protein sorting and (presumably) lipid sorting occur by lateral segregation in the plane of the membrane. The little we know about the relative membrane flux through the various pathways is reflected by the *thickness of the arrows*

cell surface displayed a shorter lag period than that of SM (Karrenbauer et al. 1990; van Meer and van't Hof 1993), which has led to the suggestion that GlcCer translocation into the lumenal leaflet (and possibly also part of its synthesis) occurs in late Golgi. Alternatively, GlcCer may be transported through the vesicular pathway more efficiently than SM (van Meer 1989).

Also concerning the vesicular nature of the transport of SM and GlcCer some uncertainty still exists, in that transport of their C_6-NBD-analogues to the cell surface remained unchanged in the presence of BFA (where transport of all proteins studied was completely blocked) (van Meer and van't Hof 1993) and, under some conditions, in mitotic cells (A.L.B. van Helvoort, R.N. Collins, and G. van Meer in prep.) in contrast with earlier findings by Kobayashi and Pagano (1989). Thus independent assays are necessary to establish the mechanism of this transport.

How and where GlcCer and galactosylceramide (GalCer) after synthesis on the cytosolic surface of the Golgi reach the lumenal surface of the vesicular transport pathway is at present unclear. A translocator seems required because spontaneous flip-flop rates for glycosphingolipids (and SM) are low (Devaux 1992). The lumenal synthesis in the Golgi of SM and the higher glycosphingolipids confines them to the exoplasmic leaflets of the various organelles. This explains why glycosphingolipids and SM after assembly in the Golgi only seem to be present in organelles linked by vesicular transport (Fig. 1) (van Meer 1989). We have established by immunoelectronmicroscopy that the complex glycosphingolipid Forssman antigen exclusively occurred in organelles that are connected by vesicular transport. It was absent from mitochondria and peroxisomes (van Genderen et al. 1991).

4 Sphingolipid Sorting in the Exocytic Pathway of Epithelial Cells

After synthesis, a variety of GlcCer analogues was preferentially delivered to the apical surface of MDCK and Caco-2 cells, with an apical/basolateral polarity of delivery between 2 and 9. In contrast, SMs arrived at the surface with polarities of typically 0.5–1 (van Meer et al. 1987; van't Hof and van Meer 1990; van't Hof et al 1992; van Meer and van't Hof 1993). Recent work has demonstrated that in MDCK cells the C_6-NBD-analogue of the glycosphingolipid GalCer, which differs from GlcCer only in the orientation of one single hydroxyl group, is preferentially transported to the basolateral domain (P. van der Bijl, I.L. van Genderen, M. Lopes-Cardozo, and G. van Meer in prep.) The simplest mechanism to generate the different polarities of delivery would be a sorting event driven by glycosphingolipid aggregation in the lumenal bilayer leaflet of the *trans*-Golgi network into an apical precursor microdomain without enrichment for SM and GalCer and under exclusion of PC. This would be followed by vesicular traffic to either plasma membrane domain (Simons and van Meer 1988; van Meer and Simons 1988). Such a mechanism is supported by the available evidence: (1) sphingolipids are synthesized in the Golgi, (2) they are transported to the cell surface by a vesicular route (although direct evidence for the translocation of GlcCer and GalCer in the Golgi is still lacking, see above), and (3) sphingolipids and not glycerolipids display a tendency to form hydrogen bonds (Pascher 1976). Transport of both GlcCer and SM to the apical cell surface, but not to the basolateral surface, was reduced by microtubule depolymerization (van Meer and van't Hof 1993). This made us conclude that they are present together in the secretory pathway before entering the apical and basolateral transport routes that originate from the trans Golgi network.

Whereas GlcCer would preferentially partition into the apical precursor domain in the *trans*-Golgi network, GalCer and SM would not. A similar difference in behaviour has been observed in the endocytic route. After endocytosis and vesicular delivery to the endosomes (Fig. 1) GlcCer traveled to the Golgi whereas SM, GalCer and LacCer recycled to the plasma membrane (Kok et al. 1991). The physicochemical basis for this behavior is at present unclear.

Analogous to the proposed self-aggregation of glycosphingolipids in the *trans*-Golgi network of epithelial cells, we have suggested that microdomain formation of the sphingo(phospho)lipid SM with cholesterol plays a part in sorting events on the *cis*-side of the Golgi (van Meer 1989).

5 Lipid Domains and Protein Sorting

We have hypothesized that microdomain formation of glycosphingolipids may be an indispensable part of the machinery that generates the lateral segregation of membrane glycoproteins in the *trans*-Golgi network of epithelial cells (Simons and van Meer 1988; van Meer and Simons 1988). The same hypothesis has been proposed for the sorting of proteins that are anchored in the membrane by a glycosylphosphatidylinositol (GPI) tail. Since in MDCK cells all GPI-proteins were found to be apical, it was proposed that they partition into the putative glucosylceramide domain (Lisanti and Rodriguez-Boulan 1990). Indeed, in FRT cells both the polarity of delivery of GPI-proteins and that of glucosylceramide were reversed (Zurzolo et al. 1993). Consistent with this notion, both GPI-proteins and sphingolipids displayed a selective resistance against extraction by Triton-X-100 in the cold (Brown 1992; Brown and Rose 1992).

6 Conclusion

The polarized distribution of lipids in epithelial cells can be explained by a lipid sorting event involving microdomain formation in the lumenal membrane leaflet of the *trans*-Golgi network. Remaining questions are: where do the simple glycosphingolipids translocate and enter the vesicular pathway? How does PC reach the cell surface of epithelial cells? Is there lipid sorting in the transcytotic route, the vesicular pathway that connects the two surface domains? What are the kinetics of metabolic conversion of cell surface lipids in relation to the kinetics of de novo synthesis and of transcytosis? Lipid domains provide an appealing scenario for sorting. However, the idea is still based on mere correlations (Brown 1992; Hannan et al. 1993). Direct proof for the occurrence of sphingolipid microdomains in cellular membranes is needed. After that, the major challenge will be to elucidate how the lumenal lipid domain is connected to cytosolic determinants of the sorting machinery, and to determine its role in intracellular lipid and protein sorting.

Acknowledgments. The present work was supported by the Netherlands Foundation for Chemical research (SON) with financial aid from the Netherlands Organization for Scientific Research (NWO) to A.v.H.

References

Brown DA (1992) Interactions between GPI-anchored proteins and membrane lipids. Trends Cell Biol 2:338–343

Brown DA, Rose JK (1992) Sorting of GPI-anchored proteins to glycolipid-enriched membrane subdomains during transport to the apical cell surface. Cell 68:533–544

Brüning A, Karrenbauer A, Schnabel E, Wieland FT (1992) Brefeldin A-induced increase of sphingomyelin synthesis. J Biol Chem 267:5052–5055

Collins RN, Warren G (1992) Sphingolipid transport in mitotic HeLa cells. J Biol Chem 267:24906–24911

Coste H, Martel MB, Got R (1986) Topology of glucosylceramide synthesis in Golgi membranes from porcine submaxillary glands. Biochim Biophys Acta 858:6–12

Devaux PF (1992) Protein involvement in transmembrane lipid asymmetry. Annu Rev Biophys Biomol Struct 31:417–439

Futerman AH, Pagano RE (1991) Determination of the intracellular sites and topology of glucosylceramide synthesis in rat liver. Biochem J 280:295–302

Futerman AH, Stieger B, Hubbard AL, Pagano RE (1990) Sphingomyelin synthesis in rat liver occurs predominantly at the cis and medial cisternae of the Golgi apparatus. J Biol Chem 265:8650–8657

Hannan LA, Lisanti MP, Rodriguez-Boulan E, Edidin M (1993) Correctly sorted molecules of a GPI-anchored protein are clustered and immobile when they arrive at the apical surface of MDCK cells. J Cell Biol 120:353–358

Hatch GM, Vance DE (1992) Stimulation of sphingomyelin biosynthesis by brefeldin A and sphingomyelin breakdown by okadaic acid treatment of rat hepatocytes. J Biol Chem 267:12443–12451

Helms JB, Karrenbauer A, Wirtz KWA, Rothman JE, Wieland FT (1990) Reconstitution of steps in the constitutive secretory pathway in permeabilized cells. Secretion of glycosylated tripeptide and truncated sphingomyelin. J Biol Chem 265:20027–20032

Holmes EH (1989) Characterization and membrane organization of $\beta 1 \rightarrow 3$- and $\beta 1 \rightarrow 4$-galactosyltransferases from human colonic adenocarcinoma cell lines Colo 205 and SW403: basis for preferential synthesis of type 1 chain lacto-series carbohydrate structures. Arch Biochem Biophys 270:630–646

Iber H, van Echten G, Sandhoff K (1992) Fractionation of primary cultured cerebellar neurons: distribution of sialyltransferases involved in ganglioside biosynthesis. J Neurochem 58:1533–1537

Jeckel D, Karrenbauer A, Birk R, Schmidt RR, Wieland F (1990) Sphingomyelin is synthesized in the *cis*-Golgi. FEBS Lett 261:155–157

Jeckel D, Karrenbauer A, Burger KNJ, van Meer G, Wieland F (1992) Glucosylceramide is synthesized at the cytosolic surface of various Golgi subfractions. J Cell Biol 117:259–267

Kallen K-J, Quinn P, Allan D (1993) Effects of brefeldin A on sphingomyelin transport and lipid synthesis in BHK21 cells. Biochem J 289:307–312

Karrenbauer A, Jeckel D, Just W, Birk R, Schmidt RR, Rothman JE, Wieland FT (1990) The rate of bulk flow from the Golgi to the plasma membrane. Cell 63:259–267

Klausner RD, Donaldson JG, Lippincott-Schwartz J (1992) Brefeldin A: insights into the control of membrane traffic and organelle structure. J Cell Biol 116:1071–1080

Kobayashi T, Pagano RE (1989) Lipid transport during mitosis. Alternative pathways for delivery of newly synthesized lipids to the cell surface. J Biol Chem 264:5966–5973

Kok JW, Babia T, Hoekstra D (1991) Sorting of sphingolipids in the endocytic pathway of HT29 cells. J Cell Biol 114:231–239

Lipsky NG, Pagano RE (1985) Intracellular translocation of fluorescent sphingolipids in cultured fibroblasts: endogenously synthesized sphingomyelin and glucocerebroside analogues pass through the Golgi apparatus en route to the plasma membrane. J Cell Biol 100:27–34

Lisanti MP, Rodriguez-Boulan E (1990) Glycophospholipid membrane anchoring provides clues to the mechanism of protein sorting in polarized epithelical cells. TIBS 15:113–118

Miller-Podraza H, Fishman PH (1982) Translocation of newly synthesized gangliosides to the cell surface. Biochemistry 21:3265–3270

Miller-Podraza H, Fishman PH (1984) Effect of drugs and temperature on biosynthesis and transport of glycosphingolipids in cultured neurotumor cells. Biochim Biophys Acta 804:44–51

Pascher I (1976) Molecular arrangements in sphingolipids. Conformation and hydrogen bonding of ceramide and their implication on membrane stability and permeability. Biochim Biophys Acta 455:433–451

Sherwood AL, Holmes EH (1992) Brefeldin A induced inhibition of de novo globo- and neolactoseries glycolipid core chain biosynthesis in human cells. J Biol Chem 267:25328–25336

Shur B (1989) Expression and function of cell surface galactosyltransferase. Biochim Biophys Acta 988:389–409

Simons K, van Meer G (1988) Lipid sorting in epithelial cells. Biochemistry 27:6197–6202

Strous GJ, van Kerkhof P, van Meer G, Rijnboutt S, Stoorvogel W (1993) Differential effects of brefeldin A on transport of secretory and lysosomal proteins. J Biol Chem 268:2341–2347

Thyberg J, Moskalewski S (1992) Reorganization of the Golgi complex in association with mitosis: redistribution of mannosidase II to the endoplasmic reticulum and effects of brefeldin A. J Submicrosc Cytol Pathol 24:495–508

Trinchera M, Fabbri M, Ghidoni R (1991) Topography of glycosyltransferases involved in the initial glycosylations of gangliosides. J Biol Chem 266:20907–20912

Trinchera M, Ghidoni R (1989) Two glycosphingolipids sialyltransferases are localized in different sub-Golgi compartments in rat liver. J Biol Chem 264:15766–15769

Trinchera M, Pirovano B, Ghidoni R (1990) Sub-Golgi distribution in rat liver of CMP-NeuAc GM3- and CMP-NeuAc:GT1bα2$\rightarrow$8 sialyltransferases and comparison with the distribution of the other glycosyltransferase activity involved in ganglioside biosynthesis. J Biol Chem 265:18242–18247

van't Hof W, Silvius J, Wieland F, van Meer G (1992) Epithelial sphingolipid sorting allows for extensive variation of the fatty acyl chain and the sphingosine backbone. Biochem J 283:913–917

van't Hof W, van Meer G (1990) Generation of lipid polarity in intestinal epithelial (Caco-2) cells: sphingolipid synthesis in the Golgi complex and sorting before vesicular traffic to the plasma membrane. J Cell Biol 111:977–986

van Echten G, Iber H, Stotz H, Takatsuki A, Sandhoff K (1990) Uncoupling of ganglioside biosynthesis by Brefeldin A. Eur J Cell Biol 51:135–139

van Genderen IL, van Meer G, Slot JW, Geuze HJ, Voorhout WF (1991) Subcellular localization of Forssman glycolipid in epithelial MDCK cells by immuno-electronmicroscopy after freeze-substitution. J Cell Biol 115:1009–1019

van Helvoort ALB, van't Hof W, Ritsema T, Sandra A, van Meer G (1994) Conversion of diacylglycerol to phosphatidylcholine on the basolateral surface of epithelial (MDCK) cells. Evidence for the reserve action of the sphingomyelin synthase. J Biol Chem (in press) van Meer G (1989) Lipid traffic in animal cells. Annu Rev Cell Biol 5:247–275

van Meer G, Burger KNJ (1992) (Glyco)sphingolipid traffic sorted out? Trends Cell Biol 2:332–337

van Meer G, Simons K (1986) The function of tight junctions in maintaining differences in lipid composition between the apical and the basolateral cell surface domains of MDCK cells. EMBO J 5:1455–1464

van Meer G, Simons K (1988) Lipid polarity and sorting in epithelial cells. J Cell Biochem 36:51–58

van Meer G, Stelzer EHK, Wijnaendts-van-Resandt RW, Simons K (1987) Sorting of sphingolipids in epithelial (Madin-Darby canine kidney) cells. J Cell Biol 105:1623–1635

van Meer G, van't Hof W (1993) Epithelial sphingolipid sorting is insensitive to reorganization of the Golgi by nocodazole, but is abolished by monensin in MDCK cells and by brefeldin A in Caco-2 cells. J Cell Sci 104:833–842

Young WW Jr, Lutz MS, Blackburn WA (1992) Endogenous glycosphingolipids move to the cell surface at a rate consistent with bulk flow estimates. J Biol Chem 267:12011–12015

Young WW Jr, Lutz MS, Mills SE, Lechler-Osborn S (1990) Use of brefeldin A to define sites of glycosphingolipid synthesis: GA2/GM2/GD2 synthase is *trans* to the brefeldin A block. Proc Natl Acad Sci USA 87:6838–6842

Zurzolo C, van't Hof W, Lisanti M, Caras I, Nitsch L, van Meer G, Rodriguez Boulan E (1993) Basolateral targeting of GPI-anchored proteins and glycosphingolipids in a polarized thyroid epithelial cell line. J Cell Biochem 178:278

Topology and Regulation of Ganglioside Metabolism – Function and Pathobiochemistry of Sphingolipid Activator Proteins

K. Sandhoff and G. van Echten[1]

1 Biosynthesis of Glycosphingolipids: Topology and Regulation

Glycosphingolipids (GSL) are amphiphilic plasma membrane components characteristic of vertebrate tissues (Svennerholm 1984; Ledeen and Yu 1982; van Echten and Sandhoff 1989).

Our current knowledge of GSL biosynthesis and their intracellular traffic has been derived from metabolic studies. The mechanism by which they are transported from sites of synthesis to other membranes has not been defined so far. It was generally assumed (Fishman and Brady 1976; Morré et al. 1979) that GSL formation is coupled, like the formation of glycoproteins, to a vesicular membrane flow, from the ER through the cisternae of the Golgi complex to the plasma membrane (for review see Schwarzmann and Sandhoff 1990). However, the involvement of glycolipid binding and/or transfer proteins in the transport of GSL cannot be excluded as long as the function of several such proteins has not been identified (Thompson et al. 1986; Tiemeyer et al. 1989).

Glycosphingolipid biosynthesis starts with the condensation of serine and palmitoyl-CoA (Fig. 1), which is catalyzed by the pyridoxal phosphate-dependent serine plamitoyltransferase (SPT), yielding 3-ketosphinganine (Mandon et al. 1991). This ketone is afterwards rapidly reduced to sphinganine by an NADPH-dependent reductase (Stoffel et al. 1968). The introduction of the 4-*trans* double bond occurs after addition of amide-linked fatty acid (Fig. 1). This sequence was proposed previously (Ong and Brady 1973; Merrill and Wang 1986; Stoffel and Bister 1974) and demonstrated recently by Rother et al. (1992). Thus sphingosine is not an intermediate of the de novo biosynthetic pathway.

All enzymatic steps involved in biosynthesis of ceramide appear to be located on the cytosolic face of the endoplasmic reticulum (Mandon et al. 1992).

The next step in the biosynthesis of GSL and sialylated GSL (gangliosides), the glucosylation of ceramide, is localized in the Golgi apparatus. Coste et al. 1986, and very recently Trinchera et al. 1991, showed that ceramide glucosyltransferase is accessible from the cytoplasmic side of Golgi vesicles. Also the following step, formation of lactosylceramide (LacCer), a common precursor of most glycosphingolipid families, was reported to occur on the cytosolic side of the Golgi apparatus (Trinchera et al. 1991). However, the finding that mutant CHO cells, with an intact galactosyltransferase I (LacCer-synthase) but lacking the translocator for UDP-Gal into the

[1] Institut für Organische Chemie und Biochemie der Universität Bonn,
Gerhard-Domagk-Straße 1, D-53121 Bonn 1, FRG.

44. Colloquium Mosbach 1993
Glyco- and Cellbiology

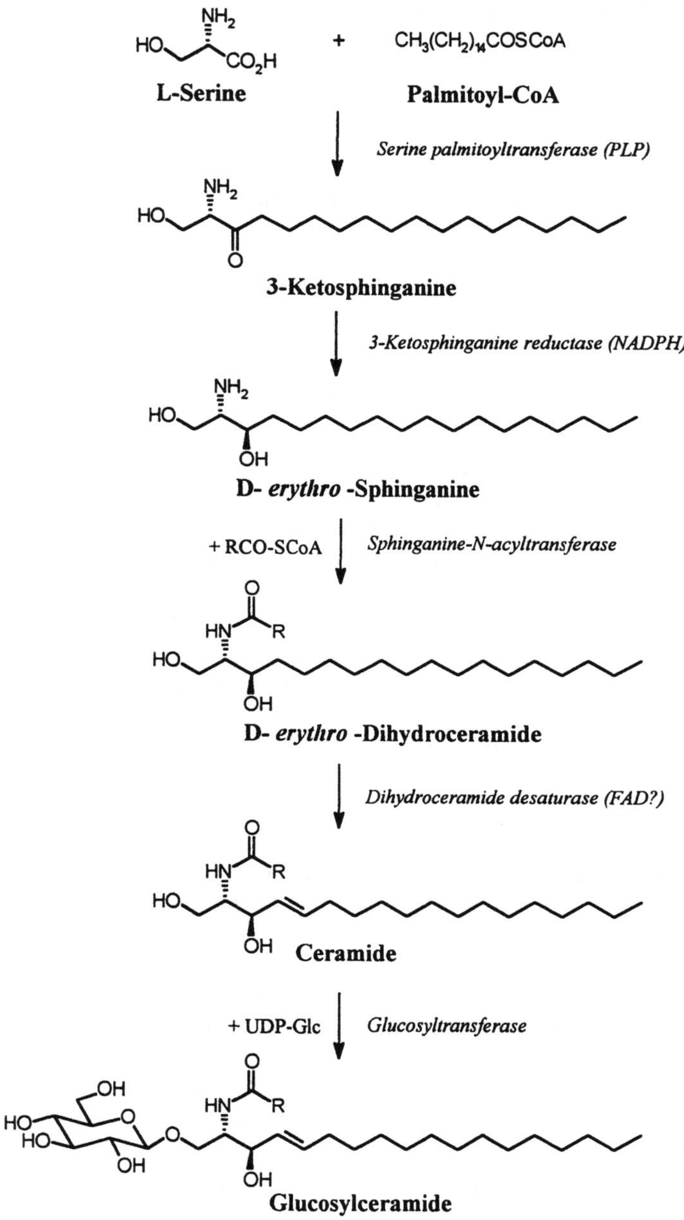

Fig. 1. Scheme for sphingolipid biosynthesis from serine to GlcCer. (Sandhoff et al. 1992)

Golgi lumen, have greatly reduced levels of LacCer (Deutscher and Hirschberg 1986) strongly suggests a luminal topology for LacCer formation.

The sequential addition of monosaccharide or sialic acid residues to the growing oligosaccharide chain, yielding GM3 and more complex gangliosides, is catalyzed by membrane-bound glycosyltransferases, which have been shown to be restricted to the luminal surface of the Golgi apparatus (for review see Schwarzmann and Sandhoff

1990). Therefore, a transfer of GlcCer or LacCer from the cytosolic to the luminal side of the Golgi membranes is required, but its occurrence has not been experimentally proven so far.

Recent studies demonstrated that transfer of the same sugar residue to analogous glycolipid acceptors, differing only in the number of neuraminic acid residues bound to the inner galactose of the oligosaccharide chain, is catalyzed by one and the same glycosyltransferase in rat liver Golgi (Pohlentz et al. 1988; Iber et al. 1989; Iber and Sandhoff 1989; Iber et al. 1991; Iber et al. 1992a).

It is, however, not yet clear where exactly in the Golgi stack the individual glycosylation reactions take place. Indirect evidence for the localization of individual steps of GSL-biosynthesis came from cell culture studies (van Echten and Sandhoff 1989; van Echten et al. 1990a). Subfractionation of the Golgi apparatus of cultured primary neurons supported previous findings (Iber et al. 1992b) that the first steps of ganglioside biosynthesis, the formation of GM3 and GD3, occur in the *cis*-Golgi compartment, while sialylation to more complex gangliosides like GD1a and GT1b takes place in the *trans*-Golgi and *trans*-Golgi network.

The expression of cell type- and differentiation-specific GSL pattern on cell surfaces is most likely regulated primarily at the transcriptional level of the respective glycosyltransferase genes (Hashimoto et al. 1983; Nakakuma et al. 1984; Nagai et al. 1986). However, some evidence is also available for an epigenetic regulation of GSL biosynthesis (Mandon et al. 1991; van Echten et al. 1990b; Yusuf et al. 1987; Iber et al. 1990). Recent studies indicate that sphingosine, the long chain base backbone and a degradation product of GSL, downregulates the first step of sphingolipid biosynthesis, formation of 3-ketosphinganine catalyzed by serine palmitoyltransferase (SPT) (Mandon et al. 1991; van Echten et al. 1990b).

2 Topology and Mechanism of Lysomomal Glycolipid Degradation, Role of Sphingolipid Activator Proteins

Components and fragments of the plasma membrane (PM) reach the lysosomal compartment mainly by an endocytic membrane flow through the early and late endocytic reticulum (Griffiths et al. 1988). During this vesicular membrane flow, molecules are subjected to a sorting process which directs some of the molecules to the lysosomal compartment and others to the Golgi or even back to the PM (Koval and Pagano 1989, 1990; Wessling-Resnick and Braell 1990; Kok et al. 1991). It remains, however, an open question whether components of the PM will be included as components of the lysosomal membrane after successive steps of vesicle budding and fusion along the endocytotic pathway. We think that it is quite unlikely that the components of the lysosomal membrane originating from the PM should be more or less selectively degraded by the lysosomal enzymes.

Alternatively, the observation of multivesicular bodies at the level of the early and late endosomal reticulum (Kok et al. 1991; Hopkins et al. 1990; McKanna et al. 1979) suggests that parts of the endosomal membranes – possibly those enriched in components derived from the PM – bud off into the endosomal lumen and thus form in-

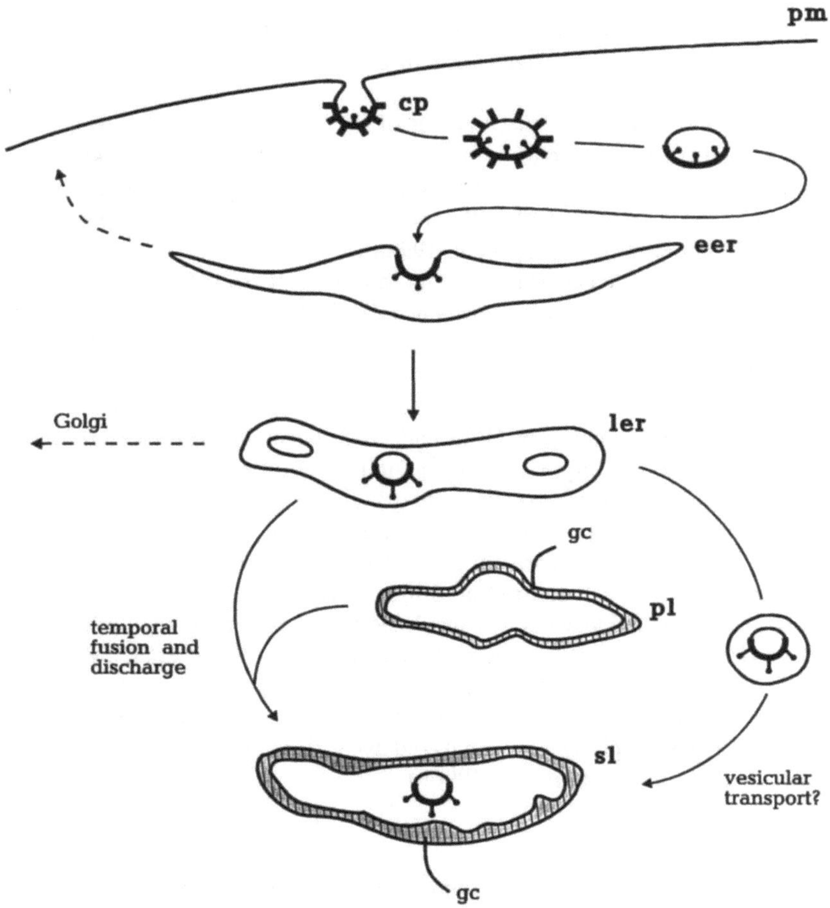

Fig. 2. A new model for the topology of endocytosis and lysosomal digestion of GSL derived from the plasma membrane. (Fürst and Sandhoff 1992). During endocytosis, glycolipids of the pm are supposed to end up in intraendosomal vesicles (multivesicular bodies), from where they are discharged into the lysosomal compartment. *pm* Plasma membrane, *cp* coated pit; *eer* early endosomal reticulum; *gc* glycocalyx; *ler* late endosomal reticulum; *pl* primary lysosome; *sl* secondary lysosome; ⚲ glycolipid. → Proposed pathway of endocytosis of GSL derived from the pm into the lysosomal compartment; – –→ other intracellular routes for GSL derived from the pm

traendosomal vesicles. These vesicles enriched in PM components could be delivered by successive processes of membrane fission and fusion along the endocytic pathways directly into the lysosol for final degradation of their components (Fig. 2). Thus glycoconjugates originating from the outer leaflet of the PM would enter the lysosol on the outer leaflet of endocytic vesicles, facing the digestive juice of the lysosomes. This hypothesis is supported by the accumulation of multivesicular storage bodies in Kupffer cells of patients with a complete deficiency of the *sap* (sphingolipid activator protein) precursor protein with a combined activator protein deficiency (Harzer et al.

1989; Schnabel et al. 1992) and by the observation that the epidermal growth factor receptor derived from the plasma membrane and internalized into lysosomes of hepatocytes is not integrated into the lysosomal membrane (Renfrew and Hubbard 1991).

Degradation of GSL occurs by stepwise action of specific acid hydrolases. Several of these enzymes need the assistance of small glycoprotein cofactors, the so-called sphingolipid activator proteins (SAPs) (Fürst and Sandhoff 1992) to attack their lipid substrates.

Since the discovery of sulfatide activator protein (Mehl and Jatzkewitz 1964), several other factors were described but their identity, specificity, and function often remained unclear. When sequence data became available, it turned out that only two genes code for the five SAPs known so far (Fürst and Sandhoff 1992). One gene carries the genetic information for the GM2-activator and the second for the *sap* precursor which is processed to four homologue proteins, including sulfatide activator protein (*sap*-B) and glucosylceramidase activator protein (*sap*-C).

Several experimental data (Meier et al. 1991) suggest the mechanism of action of GM2 activator. Hexosaminidase A is a water soluble enzyme which acts on substrates of the membrane surface only if they extend far enough into the aqueous phase (Fig. 3). Like a razor blade or a lawnmower, the enzyme recognizes and cleaves all substrates (e.g., GD1a-GalNAc) which stick out far enough into the aqueous space. However, those GSL substrates with oligosaccharide headgroups too short to be reached by the water-soluble enzyme cannot be degraded. Their degradation requires a second component, the GM2-activator, a specialized GSL binding protein, which complexes the substrate (e.g., ganglioside GM2), lifts it from the membrane, and presents it to the hexosaminidase A for degradation. While GM2-activator and hexosaminidase A represent a selective and precisely tuned machinery for the degradation of only few structurally similar sphingolipids, *sap*-B stimulates the degradation of many lipids by several enzymes from human, plant, and even bacterial origin (Li et al. 1988). Thus *sap*-B seems to act as a kind of physiological detergent with broad specificity and solubilizes glycolipid substrates (Vogel et al. 1991).

Unlike GM2-activator and *sap*-B, *sap*-C is reported to form complexes with membrane-associated enzymes and apparently activates them (Berent and Radin 1981; Ho and Light 1973; Ho 1975; Ho and Rigby 1975).

GSL with short hydrophilic headgroups are degraded by a two-component system: an enzyme and an activator protein. This seems to be much safer for the stability of short chain GSL outside the lysosomal compartment, e.g., on the cell surface. There, GSL are protected against premature degradation by lysosomal exohydrolases escaping from the cells into the extracellular fluid by low concentrations of both proteins involved in each degradation step and by an unfavourably high pH value.

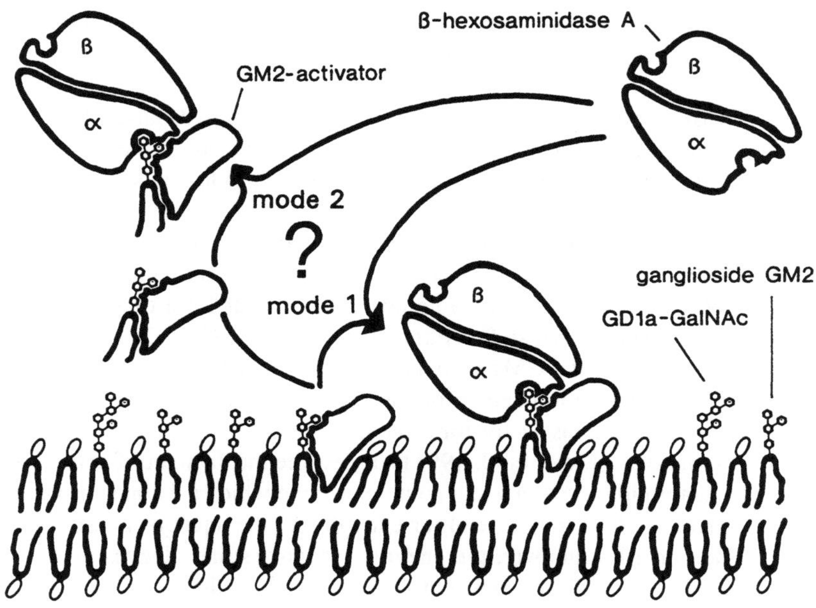

Fig. 3. Model for the GM2 activator stimulated degradation of ganglioside GM2 by human hexosaminidase A. (Fürst and Sandhoff 1992). Water-soluble hexosaminidase A does not degrade membrane-bound ganglioside GM2 in the absence of GM2 activator or appropriate detergents, but it degrades analogues of ganglioside GM2 which contain a short acyl residue or no acyl residue (lysoganglioside GM2). They are less firmly bound to the lipid bilayer and more water-soluble than GM2. Ganglioside GM2 bound to a lipid-bilayer, e.g., of an intralysosomal vesicle (see Fig. 2), is hydrolyzed in the presence of the GM2 activator. The GM2 activator binds one ganglioside GM2 molecule and lifts it a few Å out of the membrane. This activator/lipid complex can be reached and recognized by water-soluble hexosaminidase A, which cleaves the substrate. However, it is also possible that the activator/lipid complex leaves the membrane and the enzymatic reaction takes place in free solution. The terminal GalNAc residue of membrane-incorporated ganglioside GD1a-GalNAc protrudes from the membrane far enough to be accessible to hexosaminidase A without an activator

3 Pathobiochemistry of Inherited Enzyme and Activator Protein Deficiencies

As already mentioned, final degradation of sphingolipids occurs in the lysosome. Here they are degraded in a stepwise manner starting at the hydrophilic end of the molecules (Fig. 4). As indicated in Fig. 4, almost each of the degrading hydrolases can be deficient in a human lipid storage disease. The inherited deficiency of one of these ubiquitously occuring enzymes causes the lysosomal storage of its substrates. The diseases resulting from these defects are rather heterogenous from the biochemical as well as from the clinical point of view (for review see Moser et al. 1989; Barranger and Ginns 1989; O'Brien 1989; Sandhoff et al. 1989).

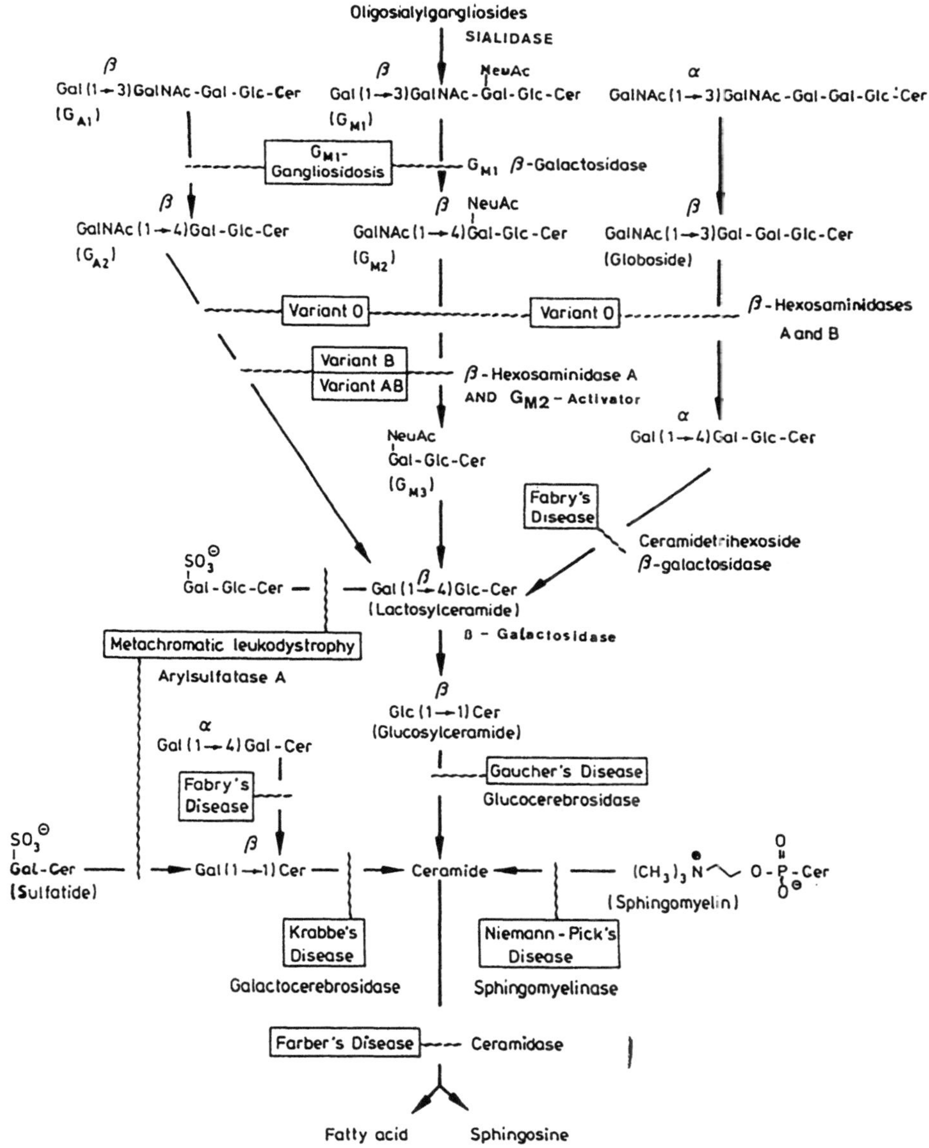

Fig. 4. Degradation scheme of sphingolipids denoting metabolic blocks of known diseases. (Sandhoff and Christomanou 1979). *Variant B* of infantile GM2-gangliosidosis, Tay-Sachs disease; *variant O* of GM2-gangliosidosis, hexosamindase β-subunit deficiency, Sandhoff disease; *variant AB* of infantile GM2-gangliosidosis

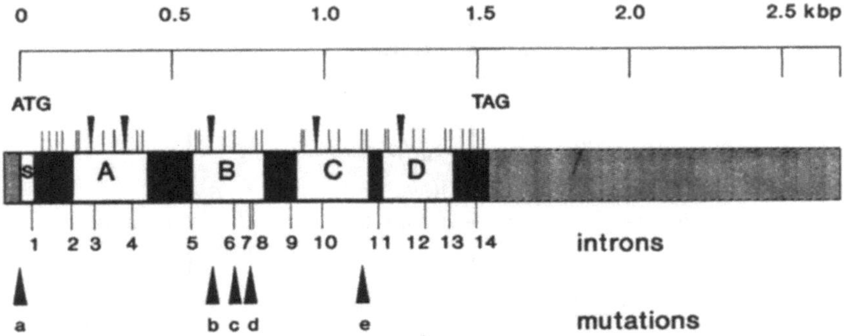

Fig. 5. Structure of the *sap*-precursor cDNA. (Fürst and Sandhoff 1992). The cDNA of *sap*-precursor codes for a sequence of 524 amino acids (or of 527 amino acids, see Holtschmidt et al. 1991), including a signal peptide of 16 amino acids (termed *s*, for the entry into the ER) (Nakano et al. 1989; Fürst and Sandhoff 1992). The four domains on the precursor, termed saposins *A-D* by O'Brien et al. (1988), correspond to the mature proteins found in human tissues. *A sap*-A or saposin A; *B sap*-B or saposin B or SAP-1 or sulfatide activator; *C sap*-C or SAP-2 or saposin C or glucosylceramidase activator protein, *D sap*-D or saposin D or component C. The positions of cysteine residues are marked by *vertical bars* and those of the N-glycosylation sites by *arrowheads*. The positionss of the 14 introns and of the known mutations leading to diseases are also given. *a* A1 → T (Met1 → Leu). (Schnabel et al. 1992); *b* C650 → T (Thr 217 → Ile). (Rafi et al. 1990; Kretz et al. 1990); *c* 33-bp insertion after G777 (11 additional amino acids after Met 259). (Zhang et al. 1990; 1991); *d* G722 → C (Cys241 → Ser). (Holtschmidt et al. 1991); *e* G1154 → T (Cys 385 → Phe). (Schnabel et al. 1991)

The analysis of sphingolipid storage diseases without detectable hydrolase deficiency resulted in the identification of several point mutations in the GM2-activator gene (reviewed by Fürst and Sandhoff 1992) and in the *sap*-precursor gene (Fig. 5). Interestingly enough, mutations affecting the *sap*-C domain (Gaucher factor) resulted in a variant form of Gaucher disease, mutations affecting the *sap*-B domain (sulfatide activator) resulted in variant forms of metachromatic leukodystrophy, whereas a mutation in the start codon ATG of the *sap*-precursor resulted in a defect of several *saps* and in a simultaneous storage of ceramide, glucosylceramide, lactosylceramide, and ganglioside GM3 in the patient's tissue.

References

Barranger YA, Ginns EJ (1989) Glucosylceramide lipidosis: Gaucher Disease. In: Scriver CR, Beaudet AL, Sly WS, Valle D (eds) The metabolic bases of inherited disease. 6th ed. vol. II, Mc Graw-Hill, New York, pp 1677–1698

Berent SL, Radin NS (1981) Mechanism of activation of glucocerebrosidase by co-β-glucosidase (glucosidase activator protein). Biochim Biophys Acta 664: 572–582

O'Brien JS, Kretz KA, Dewji N, Wenger DA, Esch F, Fluharty AL (1988) Coding of two sphingolipid activator proteins (SAP-1 and SAP-2) by same genetic locus. Science 241:1098–1101

O'Brien JS (1989) β-Galactosidase deficiency (GM1, gangliosidosis, galactosialidosis, and Morquio Syndrome Type B); ganglioside sialidase deficiency (Mucolipidosis (IV). In: Scri-

ver CR, Beaudet AL, Sly WS, Valle D (eds) The metabolic bases of inherited disease 6th ed., vol. II., Mc Graw-Hill, New York, pp 1797–1806
Coste H, Martel M-B, Got R (1986) Topology of glusosylceramide synthesis in Golgi membranes from porcine submaxillary glands. Biochim Biophys Acta 858:6–12
Deutscher SL, Hirschberg CB (1986) Mechanism of galactosylation in the Golgi apparatus. J Biol Chem 261:96–100
Fishman PH, Brady RO (1976) Biosynthesis and function of gangliosides. Science 194:906–915
Fürst W, Sandhoff K (1992) Activator proteins and topology of lysosomal sphingolipid catabolism. Biochim Biophys Acta 1126:1–16
Griffiths G, Hoflack B, Simons K, Mellman I, Kornfeld S (1988) The mannose-6-phosphate receptor and the biogenesis of lysosomes. Cell 52:329–341
Hashimoto Y, Suzuki A, Yamakawa T, Miyashita N, Moriwaki K (1983) Expression of GM1 and GD1a in mouse liver is linked to the H_2 complex on chromosome 17. J Biochem 94:2043–2048
Ho MW, Light ND (1973) Glucocerebrosidase: reconstitution from macromolecular components depends on acidic phospholipids. Biochem J 136:821–823
Ho MW (1975) Specificity of low molecular weight glycoprotein effector of lipid glycosidase. FEBS Lett 53:243–247
Ho MW, Rigby M (1975) Glucocerebrosidase. Stoichiometry of association between effector and catalytic proteins. Biochim Biophys Acta 397:267–273
Holtschmidt H, Sandhoff K, Kwon HY, Harzer K, Nakano T, Suzuki K (1991) Sulfatide activator protein: alternative splicing generates three mRNAs and a newly found mutation responsible for a clinical disease. J Biol Chem 266:7556–7560
Hopkins CR, Gibson A, Shipman M, Miller K (1990) Movement of internalized ligand-receptor complexes along a continuous endosomal reticulum. Nature 346:335–339
Harzer K, Paton BC, Poulos A, Kustermann-Kuhn B, Roggendorf W, Grisar T, Popp M (1989) Sphingolipid activator protein (SAP) deficiency in a 16-week-old atypical Gaucher disease patient and his fetal sibling: biochemical signs of combined sphingolipidoses. Eur J Pediatr 149:31–39
Iber H, Kaufmann R, Pohlentz G, Schwarzmann G, Sandhoff K (1989) Identity of GA1-, GM1a- and GD1b synthase in Golgi vesicles from rat liver. FEBS Lett 248:18–22
Iber H, Sandhoff K (1989) Identity of GD1c, GT1a and GQ1b synthase in Golgi vesicles from rat liver. FEBS Lett 254:124–128
Iber H, van Echten G, Klein RA, Sandhoff K (1990) pH-Dependent changes of ganglioside biosynthesis in neuronal cell culture. Eur J Cell Biol 52:236–240
Iber H, van Echten G, Sandhoff K (1991) Substrate specificity of α 2 $\rightarrow$ 3 sialyltransferases in ganglioside biosynthesis of rat liver Golgi. Eur J Biochem 195:115–120
Iber H, Zacharias C, Sandhoff K (1992a) The c-series gangliosides GT3, GT2 and GP1c are formed in rat liver Golgi by the same set of glycosyltransferases that catalyze the biosynthesis of asialo, a-, and b-series gangliosides. Glycobiology 2:137–142
Iber H, van Echten G, Sandhoff K (1992b) Fractionation of primary cultured neurons: distribution of sialyltransferases involved in ganglioside biosynthesis. J Neurochem 58:1533–1537
Kok JW, Babia T, Hoekstra D (1991) Sorting of sphingolipids in the endocytic pathway of HT 29 cells. J Cell Biol 114:231–239
Koval M, Pagano RE (1989) Lipid recycling between the plasma membrane and intracellular compartments: transport and metabolism of fluorescent sphingomyelin analogs in cultured fibroblasts. J Cell Biol 108:2169–2181
Koval M, Pagano RE (1990) Sorting of an internalized plasma membrane lipid between recycling and degradative pathways in normal and Niemann-Pick, type A fibroblasts. J Cell Biol 111:429–442
Kretz KA, Carson GS, Morimoto S, Kishimoto Y, Fluharty AL, O'Brien JS (1990) Characterization of a mutation in a family with saposin B deficiency: a glucosylation site defect. Proc Natl Acad Sci USA 87:2541–2544

Ledeen RW, Yu RK (1982) New strategies for detection and resolution of minor gangliosides as applied to brain fucogangliosides. Methods Enzymol 83:139–189
Li S-C, Sonnino S, Tettamanti G, Li Y-T (1988) Characterization of a nonspecific activator protein for the enzymatic hydrolysis of glycolipids. J Biol Chem 263:6588–6591
Mandon EC, van Echten G, Birk R, Schmidt RR, Sandhoff K (1991) Sphingolipid biosynthesis in cultured neurons. Downregulation of serine palmitoyltransferase by sphingoid bases. Eur J Biochem 198:667–674
Mandon E, Ehses I, Rother J, van Echten G, Sandhoff K (1992) Subcellular localization and membrane topology of serine palmitoyltransferase, 3-dehydrosphinganine reductase and sphinganine N-acyltransferase in mouse liver. J Biol Chem 267:11144–11148
McKanna JA, Haigler HT, Cohen S (1979) Hormone receptor topology and dynamics: morphological analysis using ferritin-labeled epidermal growth factor. Proc Natl Acad Sci USA 76:5689–5693
Mehl E, Jatzkewitz H (1964) Eine Cerebrosidsulfatase aus Schweineniere. Hoppe-Seyler's Z Physiol Chem 339:260–276
Meier EM, Schwarzmann G, Fürst W, Sandhoff K (1991) The human GM2 activator protein: a substrate-specific cofactor of hexosaminidase A. J Biol Chem 266:1879–1887
Merrill AH, Wang E (1986) Biosynthesis of long-chain (sphingoid) bases from serine by LM cells. J Biol Chem 261:3764–3769
Morré DJ, Kartenbeck J, Franke WW (1979) Membrane flow and interconversions among endomembranes. Biochim Biophys Acta 559:71–152
Moser W, Moser AB, Chen WW, Schram AW (1989) Ceramidase deficiency: Farber lipogranulomatosis. In: Scriver CR, Beaudet AL, Sly WS, Valle D (eds) The metabolic bases of inherited disease, 6th ed., vol. II. Mc Graw-Hill, New York, pp 1645–1654
Nagai Y, Nakaishi H, Sanai Y (1986) Gene transfer as a novel approach to the gene-controlled mechanism of the cellular expression of glycosphingolipids. Chem Phys Lipids 42:91–103
Nakakuma H, Sanai Y, Shiroki K, Nagai Y (1984) Gene-regulated expression of glycolipids: appearance of GD3 ganglioside in rat cells on transfection with transforming gene E 1 of human adenovirus type 12 DNA and its transcriptional subunits. J Biochem 96:1471–1480
Nakano T, Sandhoff K, Stümper J, Chrisomanou H, Suzuki K (1989) Structure of full-length cDNA coding for sulfatide activator, a co-β-glucosidase and two other homologous proteins: two alternate forms of the sulfatide activator. J Biochem (Tokyo) 105:152–154
Ong DE, Brady RN (1973) In vivo studies on the introduction of the 4-t-double bond of the sphingenene moiety of rat brain ceramides. J Biol Chem 248:3884–3888
Pohlentz G, Klein D, Schwarzmann G, Schmitz D, Sandhoff K (1988) Both GA2-, GM2, and GD2 synthases and GM1b, GD1a and GT1b synthases are single enzymes in Golgi vesicles from rat liver. Proc Natl Sci USA 85:7044–7048
Rafi MA, Zhang X-L, De Gala G, Wenger DA (1990) Detection of a point mutation in sphingolipid activator protein-1 mRNA in patients with a variant form of metachromatic leukodystrophy. Biochem Biophys Res Commun 166:1017–1023
Renfrew CA, Hubbard AL (1991) Degradation of epidermal growth factor receptor in rat liver. Membrane topology through the lysosomal pathway. J Biol Chem 266:21265–21273
Rother J, van Echten G, Schwarzmann G, Sandhoff K (1992) Biosynthesis of sphingolipids: dihydroceramide and not sphinganine is desaturated by cultured cells. Biochem Biophys Res Commun 189:14–20
Sandhoff K, Christomanou H (1979) Biochemistry and genetics of gangliosidoses. Hum Genet 50:107–143
Sandhoff K, Conzelmann E, Neufeld ET, Kaback MM, Suzuki K (1989) The GM2-gangliosidosis. In: Scriver CR, Beaudet AL, Sly WS, Valle D (eds) The metabolic bases of inherited disease, 6th ed., vol. II. Mc Graw-Hill, New York, pp 1807–1839
Sandhoff K, van Echten G, Schröder M, Schnabel D, Suzuki K (1992) Metabolism of glycolipids. The role of glycolipid-binding proteins in the function and pathobiochemistry of lysosomes. Biochem Soc Trans 20:695–699.
Schnabel D, Schröder M, Sandhoff K (1991) Mutation in the sphingolipid activator protein 2 in a patient with a variant of Gaucher disease. FEBS Lett 284:57–59

Schnabel D, Schröder M, Fürst W, Klein A, Hurwitz R, Zenk T, Weber G, Harzer K, Paton B, Poulos A, Suzuki K, Sandhoff K (1992) Simultaneous deficiency of sphingolipid activator proteins 1 and 2 is caused by a mutation in the initiation codon of their common gene. J Biol Chem 267:3312–3315

Schwarzmann G, Sandhoff K (1990) Metabolism and intracellular transport of glycosphingolipids. Perspectives in biochemistry, Biochemistry 29:10865–10871

Stoffel W, Le Kim D, Sticht G (1968) Metabolism of sphingosine bases: biosynthesis of dihydrosphingosine in vitro. Hoppe-Seyler's Z Physiol Chem 349:664–670

Stoffel W, Bister K (1974) Desaturation of sphinganine. Ceramide and sphingomyelin metabolism in the rat and in BHK 21 cells in tissue culture. Hoppe-Seyler's Z Physiol Chem 355:911–923

Svennerholm L (1984) Biological significance of gangliosides. In: Dreyfus H, Massarelli R, Freysz L, Rebel G (eds) Cellular and pathological aspects of glycoconjugate metabolism, vol 126. INSERM, France, pp 21–44

Thompson LK, Horowitz PM, Bently KL, Thomas DD, Alderete JF, Klebe RJ (1986) Localization of the ganglioside-binding site of fibronectin. J Biol Chem 261:5209–5214

Tiemeyer M, Yasuda Y, Schnaar RL (1989) Ganglioside-specific binding protein on rat brain membranes. J Biol Chem 264:1671–1681

Trinchera M, Fabbri M, Ghidoni R (1991) Topography of glycosyltransferases involved in the initial glycosylations of gangliosides. J Biol Chem 266:20907–20912

van Echten G, Sandhoff K (1989) Modulation of ganglioside biosynthesis in primary cultured neurons. J Neurochem 52:207–214

van Echten G, Iber H, Stotz H, Takatsuki A, Sandhoff K (1990a) Uncoupling of ganglioside biosynthesis by Brefeldin A. Eur J Cell Biol 51:135–139

van Echten G, Birk R, Brenner-Weiss G, Schmidt RR, Sandhoff K (1990b) Modulation of sphingolipid biosynthesis in primary cultured neurons by long-chain bases. J Biol Chem 265:9333–9339

Vogel A, Schwarzmann G, Sandhoff K (1991) Glycosphingolipid specificity of the human sulfatide activator protein. Eur J Biochem 200:591–597

Wessling-Resnick M, Braell WA (1990) The sorting and segregation mechanism of the endocytic pathway is functional in a cell-free system. J Biol Chem 265:690–699

Yusuf HKM, Schwarzmann G, Pohlentz G, Sandhoff K (1987) Oligosialogangliosides inhibit GM2- and GD3-synthesis in isolated Golgi vesicles from rat liver. Hoppe Seyler's Z Physiol Chem 368:455–462

Zhang X-L, Rafi MA, De Gala G, Wenger DA (1990) Insertion in the mRNA of a metachromatic leukodystrophy patient with sphingolipid activator protein-1 deficiency. Proc Natl Acad Sci USA 87:1426–1430

Zhang X-L, Rafi MA, De Gala G, Wenger DA (1991) The mechanism for a 33-nucleotide insertion in messenger RNA causing sphingolipid activator protein (SAP-1) – deficient metachromatic leukodystrophy. Hum Genet 87:211–215

Why Does Baker's Yeast Glycosylate Proteins?

W. Tanner[1]

1 Introduction

Glycosylated proteins are found in all eukaryotes, in many archebacteria, and exceptionally in eubacteria (Kornfeld and Kornfeld 1985; Tanner and Lehle 1987; Lechner and Wieland 1989; Messner and Sleyter 1988). The glycosylation of proteins is the most complex type of protein modification known in nature. Ten to 20% by weight of glycoproteins normally consists of saccharide moieties, and sometimes these even amount to more than 80%. Since obviously it is very costly for a cell to invest in the synthesis of these carbohydrate chains, they must have an important function. This assumption is strengthened by the observation that the initial rather evolved and complicated reaction sequence responsible for protein N-glycosylation, the dolichol cycle, has been conserved in evolution from yeast to man.

In recent years, a number of interesting results related to functional aspects of protein-bound saccharides have been reported. Leucocyte function and selectins, embryonic development and stage-specific cell surface oligosaccharides, coordination of cellular activation and the family of glycoprotein hormones are some of the relevant examples (Varki 1993). They clearly demonstrate that glycoproteins generally are involved in cellular information transfer, in communication of cells of the immunesystem with those of the endothelium, in cell-cell interaction determining in part embryogenesis, and in the crosstalk of various cells and tissues with each other via hormones. All these phenomena typically are restricted to multicellular organisms, to higher eukaryotes. The question therefore arises: why do *uni*cellular organisms, for example, baker's yeast, glycosylate proteins?

Two functional levels have to be distinguished when considering protein glycosylation. A cellular, most likely intracellular one that corresponds evolutionary speaking to the primary function, one which should be present, therefore, in *all* organisms, and secondary functions which must have arisen later during evolution; the latter are restricted to higher eukaryotes and obviously are represented by the examples referred to above. The ER-located reactions of N-glycosylation are conserved from yeast to man, *because* they must have been and still must be of importance in all organisms. Late in evolution additional, secondary functions became associated with saccharide chains, often with those parts that are synthesized late on the biosynthetic route (for example in the Golgi), i.e., by reactions which have *not* been conserved in evolution.

[1] Lehrstuhl für Zellbiologie und Pflanzenphysiologie, Universität Regensburg, D-93040 Regensburg, FRG.

44. Colloquium Mosbach 1993
Glyco- and Cellbiology

To find out why unicellular eukaryotes glycosylate proteins, might help to uncover the primary functions. Of course, yeasts (in this context mainly *Saccharomyces cerevisiae*) have a number of additional properties that are responsible for the extended studies on protein glycosylation over the years. Thus *S. cerevisiae* can be inexpensively grown in large amounts and it is thus well suited for biochemistry. The organism is extremely well characterized genetically; it can be transformed, and mutants can easily be obtained by gene disruption.

In this chapter I will try to summarize what is known about the N- and O-glycolsylation pathway of *S. cerevisiae* and discuss possible functions of protein glycosylation. Reviews on the topic have appeared (Tanner and Lehle 1987; Herscovics and Orlean 1993).

2 N-Glycosylation Pathway

Similarly to higher eukaryotes, the initial steps of the *N*-glycosylation pathway involve the stepwise assembly of the unique precursor oligosaccharide $Glc_3Man_9GlcNAc_2$ on the dolichol pyrophosphate carrier lipid and its en bloc transfer to the amide group of an Asn-Xaa-Thr/Ser sequon of nascent polypeptide chains. Subsequently, the precursor is modified in the ER and Golgi by processing glycosidases and glycosyltransferases. Whereas the initial processing is highly conserved in all eukaryotes, yeast Golgi modifications are very different from those in mammalian cells. No complex or hybrid type structures are synthesized in yeast.

The core oligosaccharide is assembled in a stepwise fashion on dolichol phosphate, and several of the reactions have been studied in detail both in vitro and in vivo (see Tanner and Lehle 1987; Herscovics and Orlean 1993).

Substantial progress has been made recently in both yeast and mammals in the characterization of the N-oligosaccharyltransferase (OTase) catalyzing the saccharide transfer to asparagine. Earlier attempts to purify this ER membrane-bound enzyme have failed due to its lability upon solubilization and, as it turns out, probably also because the transferase is an enzyme complex consisting of at least three subunits. In yeast two essential genes, *WBP1* (*w*heat germ agglutinin *b*inding *p*rotein) and *SWP1* (*s*uppresor of *wbp1*), have been isolated and characterized and shown to be essential for N-oligosaccharyl transferase activity in vivo and in vitro (Te Heesen et al. 1992; Te Heesen et al. 1993). Chemical crosslinking experiments and genetic data indicate that the two proteins form a complex. *Wbp*1p was originally purified due to its binding to WGA (Te Heesen et al. 1991) and subsequently found to be a component of the OTase. It is a mainly lumenally oriented type I ER transmembrane protein with a 32-amino-acid-long hydrophobic C-terminal anchor domain. The protein sequence predicts two potential N-glycosylation sites, which are used (Lehle and Aebi unpubl.) giving rise to a molecular mass of 48 kDa. Swp1p is 30 kDa large and also a type I transmembrane protein. Its gene was isolated as an allele specific suppressor of a *wbp1 ts*-mutant (Te Heesen et al. 1993). Since overexpression of both genes does not increase OTase activity, it was postulated that both are nonlimiting components of a larger protein complex. Using an anti-wbp1p-antibody, which is able to precipitate

OTase activity, a third 60/63-kDa component of the complex was isolated (Knauer and Lehle unpubl.). The doublet differs in the degree of glycosylation. Recently, it was shown that the mammalian oligosaccharyl transferase activity cofractionated with a complex of ribophorins I and II and a 48-kDa protein (OST48) (Kelleher et al. 1992). Sequence comparison of the yeast and mammalian components display interesting homologies. Wbp1p is 25% identical to the OST48 protein (Silberstein et al. 1992) and may be the homologous gene product. *SWP1* originally showed no significant homology to other known sequences in several databases (Te Heesen et al. 1993). However, comparing the C-terminal half of ribophorin II (MW 67 kDA) with Swbp1 (MW 30 kDa), an overall identity of 22% and a similarity of 46% was detected. Interestingly, ribophorin II also reveals two to three potential membrane spanning domains close to the C-terminus. So far, the specific function of each of the various components is not known.

3 The O-Glycosylation Pathway

Whereas the initial reactions of protein N-glycosylation proceed identically in all eukaryotic cells, as mentioned above, protein O-glycosylation differs considerably among organisms. Serine and threonine residues of proteins are glycosylated in fungal cells for example via Dol-P-Man as an intermediate which so far has not been observed in higher eukaryotes (Tanner and Lehle 1987; Babczinski and Tanner 1973). The following reaction sequence has been established for *S. cerevisiae* (Babczinski and Tanner 1973; Sharma et al. 1974):

$$\text{I} \quad \text{GDP-Man + Dol-P} \xrightleftharpoons{Mg^{2+}} \text{Dol-P-Man + GDP}$$

$$\text{II} \quad \text{Dol-P-Man + Protein(Ser/Thr)} \xrightarrow{Mg^{2+}} \text{Protein(Ser/Thr)Man + Dol-P}$$

$$\text{III} \quad \text{Protein(Ser/Thr)Man + nGDP-Man} \xrightarrow{Mn^{2+}} \text{Protein(Ser/Thr)Man}_{1+n} \text{ + nGDP.}$$

Evidence has been presented that reactions I and II proceed in the ER (Larriba et al. 1976; Haselbeck and Tanner 1983), whereas those reactions summed up in III (mannosyltransferase for α1,2 and α1,3-linked mannoses) most likely take place in the Golgi (Häusler et al. 1992), although the attachment of the second mannose in the ER cannot be excluded (Haselbeck and Tanner 1983; Kuranda and Robbins 1991). Strong in vivo evidence has been presented by Orlean (1990) that reactions I and II represent the only pathway for protein O-mannosylation in *S. cerevisiae* (see below).

The enzyme catalyzing reaction I, a membrane protein of 30 kDa, has been purified (Haselbeck and Tanner 1982; Haselbeck 1989). The corresponding gene, *DPM1*, has been cloned (Orlean et al 1988). It codes for a protein 267 amino acids in length with one potential transmembrane anchor. The protein seems to be attached to the membrane both with the C- and N-terminus. The main part of the protein either faces the lumenal or more likely the cytoplasmic side of the ER. Disruption of the *DPM1* gene resulted in a lethal phenotype (Orlean et al. 1988). The reason is not known.

Mammalian cells do not require this reaction for growth in culture (Chapman et al. 1980), which may indicate that protein O-glycosylation – a reaction sequence not present in mammalian cells – is essential in yeast.

The enzyme catalyzing reaction II has been purified (Strahl-Bolsinger and Tanner 1991; Sharma et al. 1991), and an antibody precipitating the enzyme activity was shown to react with a membrane glycoprotein of 92 kDa (Strahl-Bolsinger and Tanner 1991). With the help of an antibody affinity column followed by SDS-PAGE, the protein has been purified to homogeneity and via peptide sequences a positive clone was obtained from a genomic *S. cerevisiae* library (Strahl-Bolsinger et al. 1993). An open reading frame of 2451 bp codes for an 817 amino acid protein with three potential N-glycosylation sites. The gene has been called *PMT1* for *p*rotein *m*annosyl *t*ransferase. The hydropathy plot indicates a tripartite structure of the predicted protein: an N-terminal one third with four to six potential transmembrane helices, a C-terminal one third with probably four membrane spanning domains, and a central hydrophylic part. Since all three glycosylation sites most likely are glycosylated (Strahl-Bolsinger and Tanner 1991), the central part of the protein as well as the C-terminal end probably face the ER lumen. When the clone was expressed in *E. coli*, the bacterial extract catalyzed reaction II. Gene disruption led to a complete loss of in vitro mannosyltransfer activity from Dol-P-Man to a hexapeptide used as acceptor in the enzymatic assay. In vivo, however, protein O-mannosylation had decreased to only about 50% in the disruptant, indicating the existence of a second gene for a transferase not detectable by the enzyme assay. In the meantime, a test for the second transferase has been set up: the enzyme requires a tenfold higher concentration of peptide and its pH optimum (6.4) is about 1 pH unit lower than that for Pmt1p (M. Gentzsch unpubl.). A protein mannosyltransferase has also recently been characterized from *Candida albicans* (Weston et al. 1993).

A membrane-bound α-1,2 mannosyltransferase from *S. cerevisiae* has been purified, cloned, and sequenced (Häusler and Robbins 1992). The *MNT1* gene codes for a 41-kDa Golgi protein which is required for the attachment of the third mannosyl residue of O-linked saccharides (Häusler et al. 1992). The same gene has independently been cloned by selecting for killer toxin-resistant strains (Hill et al. 1992).

4 Functional Aspects

Saccharide moieties of glycoproteins are frequently thought to be involved in cell-cell recognition phenomena (Sharon and Lis 1993). Although this is the case, for example, in the leukocyte/endothelium interaction, altogether such examples are rare.

In *S. cerevisiae*, two mating type-specific cell surface glycoproteins are the agglutinins. Their formation is induced by the peptide pheromones, α- and a-factor, in the corresponding haploid mating partners, the a- and α-cell, respectively (Lipke and Kurjan 1992; Yanagishima 1984). The α-agglutinin is a highly N-glycosylated protein with an apparent molecular weight of >250kDa (Hauser and Tanner 1989; Terrance et al. 1987). The *Agα1* gene coding for a 68.2 kDa protein moiety has been cloned. From the DNA sequence a protein 650 amino acids long with a 19 amino acid signal peptide, with a GPI anchor and with 12 potential N-glycosylation sites is predicted (Hauser and Tanner 1989; Lipke et al. 1989). The completely deglycosylated protein still shows full biological activity (Hauser and Tanner 1989; Terrance et al. 1987), demonstrating that for mating type specific cell-cell interaction the protein moiety of α-agglutinin recognizes its a-agglutinin counter part exposed on a-cells. This interaction is caused by a 1:1 complex formation between the two agglutinins (Cappellaro et al. 1991).

The a-agglutinin consists of two parts: a small highly O-glycosylated protein with an apparent molecular weight of 18 kDa which is attached via S-S linkage(s) to a serine/threonine-rich cell wall protein potentially 725 amino acids in length (Orlean et al. 1986; Cappellaro et al. 1991; Watzele et al. 1988; Roy et al. 1991). Also the a-agglutinin retains most of its biological acitivity when deglycosylated by HF treatment (Cappellaro et al. 1991). This has recently been supported by expressing the a-agglutinin gene in *E. coli* as a fusion protein. A C-terminal 30-amino-acid-long peptide obtained from this carbohydrate-free agglutinin showed biological activity at a concentration of 6×10^{-8} M; the corresponding glycosylated peptide from authentic a-agglutinin was active when applied at 1.5×10^{-8} M (Cappellaro unpubl.). Therefore, it seems most likely that cell/cell recognition and mating type specific agglutination in *S. cerevisiae* is based mainly on a protein/protein interaction. The O-linked sugars of the a-agglutinin may only have a supportive, auxiliary function.

Thus, for baker's yeast, a major function of O- or N-linked saccharides in cell-cell interaction cannot be demonstrated.

As pointed out before, the strong conservation of the dolichol reactions of N-glycosylation certainly suggests that this protein modification has an essential function in all eukaryotes. In *S. cerevisiae*, early experiments with tunicamycin (Arnold and Tanner 1982) and with lethal *alg* mutations (Klebl et al. 1984), as well as the recent knock out experiments concerning the oligosaccharyl transferase (Te Heesen et al. 1992; Te Heesen et al. 1993), all show that protein N-glycosylation is a vital reaction. When this reaction is prevented in a growing culture, all cells complete their division cycle, daughter and mother cells separate, and they all get arrested as unbudded cells (Klebl et al. 1984). Since the nucleus in such cells proceeds through a second S-phase and stops in G2 (Vai et al. 1987), the arrested unbudded cells are "pseudo-G1" cells. Thus, during two instances of the yeast cell cycle, at bud initiation and at the G2/M transition, N-glycosylation of proteins obviously is required. The reason is fully obscure in the moment.

It has been shown conclusively that carbohydrate moieties influence the formation of oligomeric structures of invertase and acid phophatase (Esmon et al. 1987; Tammi et al 1987; Mrsa et al. 1989). These two secretory proteins exist in the cell periplasm as octamers and this form – possible simply due to its size – may be responsible for retaining the enzyme at the cell surface in association with the cell wall. It has been demonstrated repeatedly that glycosylated proteins in vitro are more stable against proteolytic degradation, and against thermal or chemical denaturation as compared to their nonglycosylated counterparts. This, as well as the lack of corresponding in vivo data, has been discussed previously (Tanner and Lehle 1987). On the other hand, the increased tendency of some proteins to aggregate when not glycosylated has been shown in vitro (Kern et al. 1992) and in vivo (Gibson et al. 1979; Hickman et al.

1977; Mizunaga et al. 1988). Due to the fact that protein folding is much in vogue nowadays, these latter effects have been taken up again and studied in more detail. Indeed, for viral proteins, it has been shown that lack of glycosylation leads to non-productive protein aggregation and to an inhibited secretion (Marquardt and Helenius 1992), confirming in a sense the old data of Gibson et al. (1979). However, is this a general phenomenon or a problem of a few specific proteins? The latter is not unlikely, since firstly the secretion of a number of glycoproteins is *not* affected when N-glycosylation is prevented (Tanner and Lehle 1987; Mizunaga et al. 1988), and secondly, if the feature discussed were a general one, then prevention of N-glycosylation should result in the same phenotype as *sec* mutations: when shifted to nonpermissive temperature *sec* mutants stop in the cell cycle wherever they are and remain a completely unsynchronized culture (Tanner et al. unpbul.). When N-glycosylation is switched off, on the other hand, young buds grow to normal-sized daughter cells, complete their cell cycle, and are highly synchronized (see above).

How about protein targeting? The mannose-6-phosphate signal pathway for lysosomal enzymes in mammalian cells is a well-documented example (Kornfeld and Mellmann 1989). In *S. cerevisiae*, however, although the mannose-6-phosphate group exists on various glycoproteins, it does not seem to be functional as a signal which targets proteins to the yeast's lysosome, its vacuole (Schwaiger et al. 1982; Valls et al. 1987; Johnson et al. 1987; Clark et al. 1982).

Whereas there is no doubt that N-glycosylation is a vital reaction for yeast, this is not yet known for O-glycosylation. Since the knock-out experiment of the Dol-P-Man: protein mannosyltransferase (see above) has disclosed the existence of a second enzyme, the answer to this question has to await the cloning of the second gene. Thus, it certainly holds for yeasts what Hakamori recently stated: "the function of glycoconjugates remains essentially a great enigma" (Hakomori 1991).

Acknowledgment. I am very grateful to all coworkers whose work is referred to in the text, especially also to Dr. Ludwig Lehle for helpful discussions and for making available unpublished data. The work from this laboratory has been supported by the Deutsche Forschungsgemeinschaft (SFB 43) and from Fonds der Chemischen Industrie.

References

Arnold E, Tanner W (1982) An obligatory role of protein glycosylation in the life cycle of yeast cells. FEBS Lett 148:49–53

Babczinski P, Tanner W (1973) Involvement of dolichol monophosphate in the formation of specific mannosyl linkages in yeast glycoproteins. Biochem Biophys Res Commun 54:1119–1124

Cappelaro C, Hauser K, Mrsa V, Watzele M, Watzele G, Gruber C, Tanner W (1991) *Saccharomyces cerivisae* a- and α-agglutinin. Characterization of their molecular interaction. EMBO J 10:4081–4088

Chapman A, Fujimoto K, Kornfeld S (1980) The primary glycosylation defect in class E Thy-1-negative mutant mouse lymphoma cells in inability to synthesize dolichol-P-mannose. J Biol Chem 255:4441–4446

Clark DW, Tkacz JS, Lampen JO (1982) Asparagine-linked carbohydrate does not determine the cellular location of yeast vacuolar nonspecific alkaline phosphatase. J Bacteriol 152:865–873

Esmon PC, Esmon BE, Schauer IE, Taylor A, Schekman R (1987) Structure, assembly, and secretion of octameric invertase. J Biol Chem 262:4387–4394

Gibson R, Schlesinger S, Kornfeld S (1979) The nonglycosylated glycoprotein of Vesicular Stomatitis Virus is temperature-sensitive and undergoes intracellular aggregation at elevated temperatures. J Biol Chem 254:3600–3607

Hakomori S (1991) All that with apparent manifest is not of primary importance. Trends Glycosci Glycotechnol 3:1–3

Haselbeck A, Tanner W (1982) Dolichyl phosphate-mediated mannosyl transfer through liposomal membranes. Proc Natl Acad Sci USA 79:1520–1524

Haselbeck A, Tanner W (1983) O-Glycosylation in *Saccharomyces cerevisiae* is initiated at the endoplasmic reticulum. FEBS Lett 158:335–338

Haselbeck A (1989) Purification of GDP mannose: dolichyl-phosphate O-β-D-mannosetransferase from *Saccharomyces cerevisiae*. Eur J Biochem 181:663–668

Hauser K, Tanner W (1989) Purification of the inducible α-agglutinin of *S. cerevisiae* and molecular cloning of the gene. FEBS Lett 255:290–294

Häusler A, Robbins PW (1992) Glycosylation in *Saccharomyces cerevisiae*: cloning and characterization of an α-1,2-mannosyltransferase structural gene. Glycobiology 2:77–84

Häusler A, Ballou L, Ballou CE, Robbins PW (1992) Yeast glycoprotein biosynthesis: MNT1 encodes an α-1,2-mannosyltransferase involved in O-glycosylation. Proc Natl Acad Sci USA 89:6846–6850

Herscovics A, Orlean P (1993) Glycoprotein biosynthesis in yeast. FASEB J 7:540–550

Hickman S, Kulczycki A Jr, Lynch RG, Kornfeld S (1977) Studies of the mechanism of tunicamycin inhibition of IgA and IgE secretion by plasma cells. J Biol Chem 252:4402–4408

Hill K, Boone C, Goebl M, Puccia R, Sdicu A-M, Bussey H (1992) Yeast KRE2 defines a new gene family encoding probable secretory proteins, and is required for the correct N-glycosylation of proteins. Genetics 130:273–283

Johnson LM, Bankaitis VA, Emr SD (1987) Distinct sequence determinants direct intracellular sorting and modification of a yeast vacuolar protease. Cell 48:875–885

Kelleher DJ, Kreibich G, Gilmore R (1992) Oligosaccharyltransferase activity is associated with a protein complex composed of ribophorins I and II and a 48 kd protein. Cell 69:55–65

Kern G, Schülke N, Schmid FX, Jaenicke R (1992) Stability, quaternary structure, and folding of internal, external, and core-glycosylated invertase from yeast. Protein Science 1:120–131

Klebl F, Huffaker TC, Tanner W (1984) A temperature-sensitive N-glycosylation mutant of *S. cerevisiae* that behaves like a cell-cycle mutant. Exp Cell Res 150:309–313

Kornfeld R, Kornfeld S (1985) Assembly of asparagine-linked oligosaccharides. Annu Rev Biochem 54:631–664

Kornfeld S, Mellman J (1989) The biogenesis of lysosomes. Annu Rev Cell Biol 5:483–525

Kuranda MJ, Robbins PW (1991) Chitinase is required for cell separation during growth of *Saccharomyces cerevisiae*. J Biol Chem 266:19758–19767

Larriba G, Elorza MV, Villanueva JR, Sentandreu R (1976) Participation of dolichol phosphomannose in the glycosylation of yeast wall mannoproteins at the polysomal level. FEBS Lett 71:316–320

Lechner J, Wieland F (1989) Structure and biosynthesis of prokaryotic glycoproteins. Annu Rev Biochem 58:173–194.

Lipke PN, Wojciechowicz D, Kurjan J (1989) AGα1 is the structural gene for the *Saccharomyces cerevisiae* α-agglutinin, a cell surface glycoprotein involved in cell-cell interactions during mating. Mol Cell Biol 9:3155–3165

Lipke PN, Kurjan J (1992) Sexual agglutination in budding yeasts: structure, function, and regulation of adhesion glycoproteins. Microbiol Rev 56:180–194

Marquardt T, Helenius A (1992) Misfolding and aggregation of newly synthesized proteins in the endoplasmic reticulum. J Cell Biol 117:505–513

Messner P, Sleyter UB (1988) Asparaginyl-rhamnose: a novel type of a protein-carbohydrate linkage in an eubacterial surface-layer glycoprotein. FEBS Lett 228:317–320

Mizunaga T, Izawa M, Ikeda K, Maruyama Y (1988) Secretion of an active nonglycosylated form of the repressible acid phosphate of *Saccharomyces cerevisiae* in the presence of tunicamycin at lower temperatures. J Biochem 103:321–326

Mrsa V, Barberic S, Ries B, Mildner P (1989) Influence of glycosylation on the oligomeric structure of yeast acid phosphatase. Arch Biochem Biophys 273:121–127

Orlean P, Ammer H, Watzele M, Tanner W (1986) Synthesis of an O-glycosylated cell surface protein induced in yeast by α factor. Proc Natl Acad Sci USA 83:6263–6266

Orlean P, Albright C, Robbins PW (1988) Cloning and sequencing of the yeast gene for dolichol phosphate mannose synthase, an essential protein. J Biol Chem 263:17499–17507

Orlean P (1990) Dolichol phosphate mannose synthase is required in vivo for glycosyl phosphatidylinositol membrane anchoring, O-mannosylation, and N-glycosylation of protein in *Saccharomcyes cerevisiae*. Mol Cell Biol 10:5796–5805

Roy A, Lu CF, Marykwas DL, Lipke PN, Kurjan J (1991) The AGA1 product is involved in cell surface attachment of the *Saccharomyces cerevisiae* cell adhesion glycoprotein a-agglutinin. Mol Cell Biol 11:4196–4206

Schwaiger H, Hasilik A, von Figura K, Wiemken A, Tanner W (1982) Carbohydrate-free carboxypeptidase Y is transferred into the lysosome-like vacuole. Biochem Biophys Res Commun 104:950–956

Sharma CB, Babczinski P, Lehle L, Tanner W (1974) The role of dolichol monophosphate in glycoprotein biosynthesis in *S. cerevisiae*. Eur J Biochem 46:35–41

Sharma CB, D'Souza C, Elbein AD (1991) Partial purification of a mannosyltransferase involved in the O-mannosylation of glycoproteins from *Saccharomyces cerevisiae*. Glycobiology 1:367–373

Sharon N, Lis H (1993) Carbohydrates in cell recognition. Scientific American 268:74–81

Silberstein S, Kelleher DJ, Gilmore R (1992) The 48-kDa subunit of the mammalian oligosaccharyltransferase complex is homologous to the essential yeast protein WBP1. J Biol Chem 267:23658–23663

Strahl-Bolsinger S, Tanner W (1991) Protein O-glycosylation in *Saccharomyces cerevisiae*. Purification and characterization of the dolichyl-phosphate-D-mannose: protein O-mannosyltransferase. Eur J Biochem 196:185–190

Strahl-Bolsinger S, Immervoll T, Deutzmann R, Tanner W (1993) PMT1, the gene for a key enzyme of protein O-glycosylation in *Saccharomyces cerevisiae*. Proc Natl Acad Sci USA 90:8164–8168

Tammi M, Ballou L, Taylor A, Ballou CE (1987) Effect of glycosylation on yeast invertase oligomer stability. J Biol Chem 262:4395-4401

Tanner W, Lehle L (1987) Protein glycosylation in yeast. Biochim Biophys Acta 906:81–99

Te Heesen S, Rauhut R, Aebersold R, Abelson J, Aebi M, Clark MW (1991) An essential 45 kDa yeast transmembrane protein reacts with anti-nuclear pore antibodies: purification of the protein, immunolocalization and cloning of the gene. Eur J Cell Biol 56:8–18

Te Heesen S, Janetzky B, Lehle L, Aebi M (1992) The yeast WBP1 is essential for oligosaccharyl transferase activity in vivo and in vitro. EMBO J 11:2071–2075

Te Heesen S, Knauer R, Lehle L, Aebi M (1993) Yeast Wbp1p and Swp1p form a protein complex essential for oligosaccharyl transferase actitivty. EMBO J 12:279–284

Terrance K, Heller P, Wu Y-S, Lipke PN (1987) Identification of glycoprotein components of α-agglutinin, a cell adhesion protein from *Saccharomyces cerevisiae*. J Bacteriol 169:475–482

Vai M, Popolo L, Alberghina L (1987) Effect of tunicamycin on cell cycle progression in budding yeast. Exp Cell Res 171:448–459

Valls LA, Hunter CP, Rothman JH, Stevens TH (1987) Protein sorting in yeast: The localization determinant of yeast vacuolar carboxypeptidase Y resides in the propeptide. Cell 48:887–897

Varki A (1993) Biological roles of oligosaccharides: all of the theories are correct. Glycobiology 3:97–130

Watzele M, Klis F, Tanner W (1988) Purification and characterization of the inducible a agglutinin of *Saccharomyces cerevisiae*. EMBO J 7:1483–1488
Weston A, Nassau PM, Henly C, Marriott MS (1993) Protein O-mannosylation in *Candida albicans*: determination of the amino acid sequences of peptide acceptors for protein O-mannosyltransferase. Eur J Biochem 215:845–849
Yanagishima N (1984) Mating systems and sexual interactions in yeast. In: Linskens HF, Heslop-Harrison J (eds) Encyclopedia of plant physiology: cellular interactions. New Series, vol 17. Springer, Berlin Heidelberg New York, pp 402–423

Glycosylation of Nuclear and Cytoplasmic Proteins Is as Abundant and as Dynamic as Phosphorylation

G. W. Hart, W. G. Kelly, M. A. Blomberg, E. P. Roquemore, L.-Y. D. Dong, L. Kreppel, T-Y. Chou, D. Snow, and K. Greis[1]

1 Introduction

While probing murine lymphocyte cell surfaces with bovine milk galactosyltransferase, we discovered a new form of protein glycosylation in which single N-acetylglucosamine monosaccharides are O-glycosidically linked to serine or threonine moieties (O-GlcNAc) (Torres and Hart 1984). Surprisingly, O-GlcNAc is highly abundant (lymphocytes have > 1.5×10^8 molecules/cell), and is exclusively localized in nucleoplasmic and cytoplasmic compartments of the cell (Kearse and Hart 1991b). Based upon probing with galactosyltransferase (Whiteheart et al. 1989), a myriad of proteins in both the nucleus and cytoplasm are modified by O-GlcNAc. O-GlcNAc has been found in eukaryotes from yeast to man (Haltiwanger et al. 1992b; Hart et al. 1989), but has not yet been detected in prokaryotes. Identified O-GlcNAc-bearing proteins are listed in Table 1. These intracellular glycoproteins have a diverse range of functions, including transcription regulatory factors, enzymes, nuclear pore proteins, cytoskeletal proteins, and viral proteins. However, all of these glycoproteins are also phosphoproteins that form reversible multimeric complexes, depending upon their phosphorylation states, thus suggesting that O-GlcNAc may play a role in mediating/modulating regulated and reversible protein subunit interactions.

Identified O-GlcNAc attachment sites (Table 2) and studies of synthetic acceptor peptides for the purified UDP-GlcNAc:polypeptide N-acetylglycosaminyltransferase (O-GlcNAc transferase) (Haltiwanger et al. 1990; Haltiwanger et al. 1992a) suggest that the sites of O-GlcNAc addition are virtually indistinguishable from sites also utilized by the "growth-factor" or "proline-specific" family of protein kinases (Ralph et al. 1990; Roach 1991; Meek and Street 1992; Taylor and Adams 1992). Thus, O-GlcNAc may in some cases have a "yen-yang" relationship with phosphorylation, perhaps by blocking site-specific phosphorylation. Several examples showing a reciprocal relationship of O-GlcNAc and phosphorylation have already been found, including the CTD on the catalytic subunit of RNA polymerase II (Kelly et al. 1993) and the cytokeratins (Chou et al. 1992a; Chou and Omary 1993). However, localization studies have not yet been carried out to evaluate the possible reciprocal occupancy of specific hydroxyl moieties in any protein. O-GlcNAc attachment sites also have a high PEST score (Rogers et al. 1986; Hirai et al. 1991). PEST domains appear to target intracellular proteins for rapid degradation. Attachment of O-GlcNAc to

[1] Department of Biochemistry and Molecular Genetics, University of Alabama at Birmingham Schools of Medicine/Dentistry, 1918 University Boulevard, Birmingham, AL 35294–0005, USA.

44. Colloquium Mosbach 1993
Glyco- and Cellbiology

Table 1. Identified O-GlcNAc proteins

Protein	Reference	Protein	Reference
Human erythrocyte band 4.1	(Holt et al. 1987)	Many schistosome proteins	(Nyame et al. 1987)
Synapsin I	(Luthi et al. 1991)	Many trypanosome proteins	(Kelly and Hart unpubl.)
Cytokeratins – 13, 8, 18	(King and Hounsell 1989; Chou et al. 1992; Chou and Omary 1993)	vErb-A oncogene protein	(Privalsky 1990)
Neurofilaments	(Dong et al. 1993)	c-Myc oncogene protein	(Chou et al. in prep.)
Talin	(Hagmann et al. 1992)	Adenovirus fiber protein	(Mullis et al. 1990)
67 kDa RBC E.F. kinase	(Datta et al. 1989)	CMV tegument protein	(Benko et al. 1988)
65 kDa nuclear tyrosine phosphatase	(Meikrantz et al. 1991)	Baculovirus tegument protein	(Whitford and Faulkner 1992)
Nuclear pore complex proteins	(Holt et al. 1987; Starr and Hanover 1990; Davis and Blobel 1987; Hanover et al. 1987; Park et al. 1987; Schindler et al. 1987)	NS26 protein of rotavirus	(Gonzalez and Burrone 1992)
RNA polymerase II catalytic subunit	(Kelly et al. 1993)	*Aplysia* neuron proteins	(Gabel et al. 1989; Ambron et al. 1989)
Many RNA polymerase II transcription factors	(Jackson and Tjian 1988; Jackson and Tjian 1989; Lichtsteiner and Schibler 1989; Reason et al. 1992)	Many chromatin proteins in *Drosophila*	(Kelly and Hart 1989
Lens α-crystallins (small heat schock proteins)	(Roquemore et al. 1992)	92 kDa smooth endoplasmic reticulum protein	(Abeijon and Hirschberg 1988)

Table 2. Sites of O-GlcNAc attachment in vivo (see Table 1 for references):

Rat nuclear pore 62 kDa (one of several sites)	...MAGGPADTSDPL...
Human erythrocyte 65 kDa cytosolic protein	...DSPVSQPSLVGSK...
Human erythrocyte band 4.1	...AQTITSETPSSTT
Bovine lens α-A-crystallin	...^{158}DIPVSREEK^{166}...
Rhesus monkey α-B-crystallin	...^{164}EEKPAVTAAPK^{174}...
Recombinant human serum response transcription factor (baculovirus expressed)	...^{269}VTNLPGTTSTIQTAPSTSTTMQV SSGPSFPITNYLAPVSASVSPSAVS SANGTVLKSTGSGPVSSGGLMQLPTS FTLMPGGAVAQQVPVQAIQVHQAPQ QASPSRDSSTDLTQTSSSGTVTLPA TIMTSSVPTT^{402}...
Rat spinal cord Neurofilament (NF-L)	...^{18}YVETPRVHISSVR^{30}... ...^{38}SAYSSYSAPVSSSLSVR^{54}...
Rat spinal cord Neurofilament (NF-M)	...^{44}GSPSTVSSSYK^{54}... ...^{427}QPSVTISSK^{435}...
Chicken gizzard talin	^{1470}ANQAIQMAXQNLVDPAXTQ^{1488a} ^{1883}GILANQLTNDYGQLAQQ^{1889a}
Calf thymus RNA polymerse II	[b]...(S/T)P(S/T)SP....TPTSPN....SPTSPT...

[a] Numbers correspond to mouse sequence.
[b] Spread throughout the C-terminal domain.

these PEST sites could serve to modulate turnover of several important regulatory proteins. Unfortunately, this intriguing, and very testable hypothesis remains to be examined.

Several studies have shown that O-GlcNAc is highly dynamic. The levels of O-GlcNAc on lymphocyte nuclear and cytoplasmic proteins change within minutes after stimulation by antigens or mitogenic agents (Kearse and Hart 1991a). Pulse-chase analyses of cytokeratins indicates that the saccharide turns over many times more rapidly than the polypeptide (Chou et al. 1992a). In addition, the levels of O-GlcNAc on cytokeratins increase drammatically in mitotically arrested cells (Chou and Omary 1993). In this instance, phosphorylation and glycosylation both increase, but each modification appears to occur on different populations of proteins. O-GlcNAc levels on several proteins, including nuclear pore glycoproteins, change with various stages of the cell-cycle (W.G. Kelly, E.P. Roquemore and G.W. Hart unpubl.). Thus, O-GlcNAc displays all of the hallmarks of a regulatory modification that may have a "yen-yang" relationship with phosphorylation. As yet, direct evidence for a regulatory role for O-GlcNAc is lacking. However, the study of intracellular glycosylation is still in its early infancy. Given O-GlcNAc's widespread abundance on many of the cell's most interesting and essential proteins, it is likely that elucidation of its functions will impact all aspects of eukaryotic biology.

2 Nuclear Pore Glycoproteins

Subcellular localization studies in rat hepatocytes demonstrated that O-GlcNAc is enriched in the nucleus, with the highest relative levels in the nuclear envelope (Holt et al. 1987b). Comparisons of Western blots using monoclonal antibodies prepared against nuclear pore proteins (nucleoporins) (Snow et al. 1987), to autoradiographs of galactosyltransferase-radiolabeled nuclear envelopes showed identical sets of proteins detected by both methods (Holt et al. 1987b). Blocking by galactose capping or removal of O-GlcNAc by hexosaminidase treatments showed that all of the approximately 20 anti-nucleoporin monoclonal antibodies have O-GlcNAc as a major part of their epitopes. Similar findings have been reported by others studying nuclear pore glycoproteins (Table 1), suggesting that the clustered O-GlcNAc moieties on these molecules are highly immunogenic. Even though the nucleoporins are often multiply glycosylated, only a few attachment sites have been identified (Table 2). One of the major nucleoporins, p62, had its cDNA cloned and sequenced (D'Onofrio et al. 1988; Starr et al. 1990). Nuclear pore reconstitution studies have shown that nucleoporins are essential for pore-mediated nuclear transport (Finlay et al. 1991; Forbes 1992). Antibodies to O-GlcNAc or the lectin, wheat germ agglutinin (WGA) block nuclear transport at the ATP-requiring step (Newmeyer and Forbes 1988; Finlay et al. 1987; Hanover 1992), but do not appear to sterically impede transport. A direct role for the saccharide in nuclear transport has not been demonstrated. It seems likely that O-GlcNAc may play a role in the reversible assembly/disassembly of the nuclear pores as the cell traverses the cell cycle. A comparison of the glycosylation/phosphorylation of nucleoporins during the cell cycle should prove illuminating.

3 Chromatin Proteins

Even though the relative concentration of O-GlcNAc appears high in nucleoporins, most of the O-GlcNAc in the cell appears to be on proteins classically referred to as chromatin (Holt and Hart 1986). Staining of isolated *Drosophila* polytene chromosomes with fluorescene-tagged WGA results in an intense fluorescence along the entire length of the chromosome in a banding pattern similar to classical chromatin dyes (Kelly and Hart 1989). Specific galactosyltransferase radiolabeling of O-GlcNAc followed by autoradiography also shows dramatic concentrations of silver grains along the length of the chromosome (Fig. 1). Both WGA staining and silver grain density are much reduced at "puffs" (active sites of gene transcription).

Virtually every RNA polymerase II transcription factor that has been carefully examined is glycosylated by O-GlcNAc (see Table 1). In fact, WGA-Sepharose chromatography has become an accepted step for purification of transcription factors (Jackson and Tjian 1989). However, it is likely that only a subpopulation of any transcription factor will have high affinity to WGA-Sepharose, depending upon the amounts, accessibility, and distribution of its O-GlcNAc moieties present at the time of its isolation from the cell. Thus, biological studies using WGA-Sepharose purified factors should be interpreted with the knowledge that only a selected subpopulation of

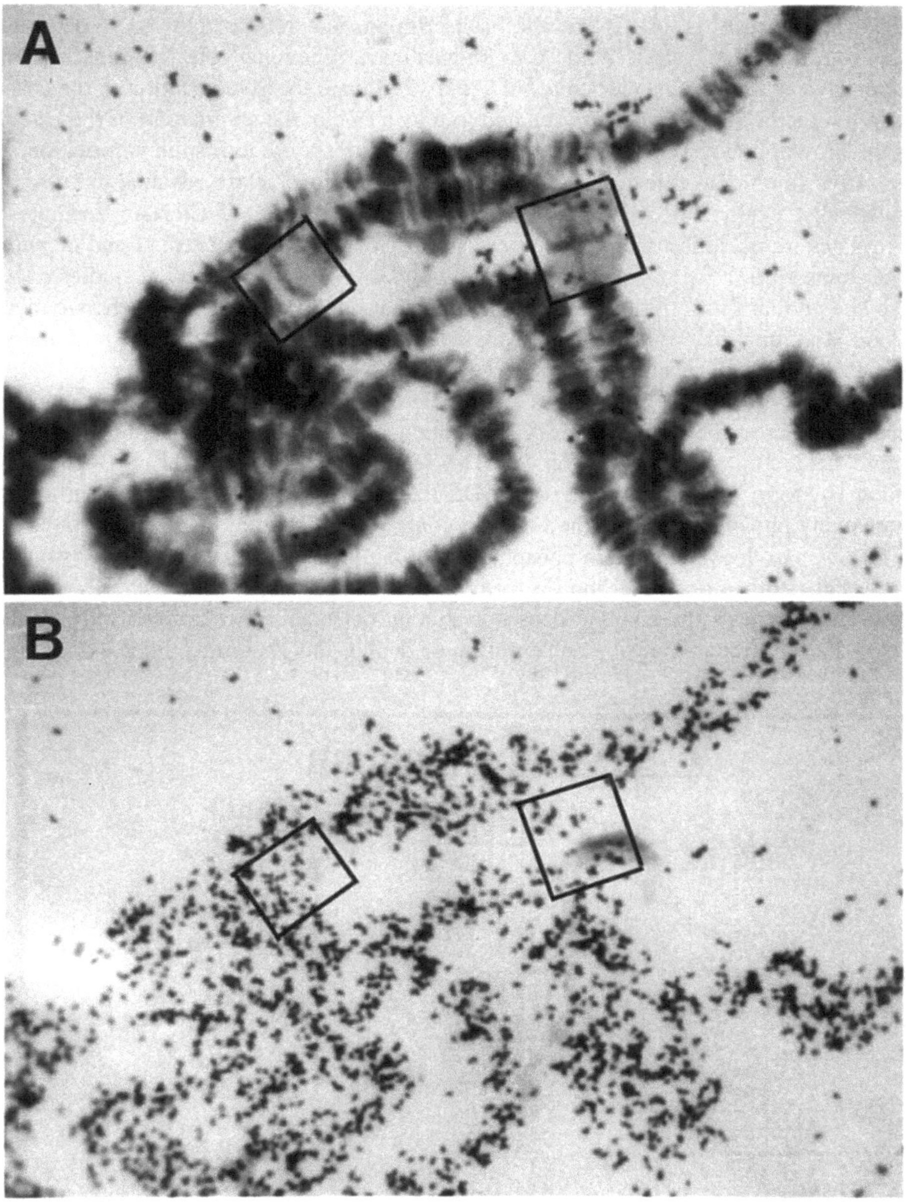

Fig. 1 A, B. Distribution of O-GlcNAc on polytene chromosomes of *Drosophilia*. Polytene chromosomes from third instar larvae fixed with formaldehyde and were radiolabeled using galactosyltransferase to detect O-GlcNAc. **A** Phase-contrast microscopy. **B** Autoradiography using photographic emulsion. "Puffed" regions (active sites of gene transcription) are *boxed*. Biochemical analyses confirm that > 90% of the radiolabel is in O-GlcNAc-bearing proteins

unknown biological selectivity is actually being studied. Thus far, glycosylation of RNA polymerase I- or III-specific transcription factors has not been detected (Jackson and Tjian 1988). While data suggesting a functional role for transcription factor glycosylation exist (Jackson and Tjian 1988), and the glycosylation of the transcription factor Sp-1 has been postulated to play a role in the stimulation of the levels of the growth factor TGFα (McClain et al. 1992) in response to insulin stimulation, a definitive functional role for O-GlcNAc on transcription factors remains to be systematically evaluated. Again, a likely hypothesis is that O-GlcNAc mediates/modulates transcription factor multimerization with RNA polymerase II and/or with other transcription factors, a notion currently under investigation. Recent studies have also shown that transcription factor-related nuclear oncogene proteins, such as c-myc (Chou et al. unpubl.) or v-Erb-A (Privalsky 1990) are also glycosylated. Preliminary studies suggest that certain steroid receptors may also contain O-GlcNAc.

All eukaryotic RNA polymerase II (Pol II) catalytic subunits contain an unusual repeat domain at their C-terminus (CTD) composed of up to 50 repeats of the seven amino acid consensus, – (Tyr-Ser-Pro-Thr-Ser-Pro-Ser) – (Corden 1990). The so-called II_O form of Pol II migrates on SDS-PAGE at ~240 kDa and is known to be extensively phosphorylated in the CTD. In contrast, the IIa form of Pol II migrates at ~215 kDa and does not contain phosphate. Probing for O-GlcNAc via glactosyltransferase has demonstrated that the IIa form of Pol II is glycosylated, but the II_O phosphorylated form of the enzyme does not contain detectable saccharide (Kelly et al. 1993). Furthermore, extensive denaturation or dephosphorylation of the II_O does not

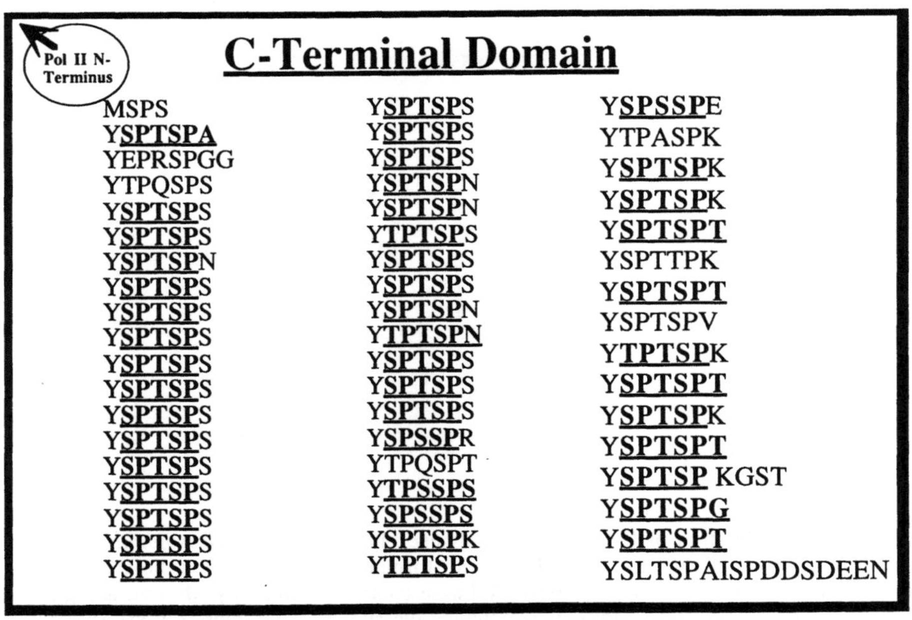

Fig. 2. Distribution of O-GlcNAc modified peptides along the C-terminal domain of RNA Polymerase II. Highlighted sequences have been shown to bear O-GlcNAc. However, precise localization and distribution remains unknown

expose O-GlcNAc moieties. Thus, phosphorylation and glycosylation of Pol II's CTD appear to be reciprocal, suggesting that the enzyme exists in the cell in at least three structurally distinct isoforms. Mapping of O-GlcNAc moieties in the CTD show that they are distributed throughout the C-terminal domain, analogous to its phosphorylation (Fig. 2 and Table 2) (Kelly et al. 1993).

Again, these date suggest that O-GlcNAc may play a role in the formation of the initiation complex of transcription. Previous studies have shown that the species of Pol II that binds to the initiation complex appears to be the nonphosphorylated (IIa) form. IIa is phosphorylated to the II_O form only after it binds to the initiation complex. Presumably, the phosphorylation event is important for allowing the elongation process to occur. These series of events suggest that a de-glycosylation event must precede the phosphorylation of the CTD. The data also allow the speculation that GlcNAc-binding proteins (lectins?) may be components of the transcription initiation complex.

4 Cytoskeletal Proteins

One of the earliest proteins shown to bear O-GlcNAc is band 4.1 of the human erythrocyte (Holt et al. 1987b). Band 4.1 is an important cytoskeletal protein that helps maintain erythrocyte shape by interacting with the cytoplasmic tail of glycophorin and the actin/spectrin cytoskeleton. In these studies, the glycosylated form of band 4.1 appeared to represent a small subset that is most tightly associated with the cytoskeleton. However, much further work is required to substantiate this observation. Talin, an important cytoskeletal protein that is localized at sites where actin filaments are linked to the plasma membrane, also is modified by O-GlcNAc (Hagmann et al. 1992). Talin interacts with the cytoplasmic domains of cell adhesion molecules, integrins, and appears to bridge the cytoskeleton to the membrane via interacting with another cytoskeletal protein, vinculin. The absence of O-GlcNAc on talin derived from platelets, in which it is not found associated with vinculin or focal plaques, led to the suggestion that the saccharide may play a role in the association of vinculin with talin (Hagmann et al. 1992).

Several different intermediate filaments contain O-GlcNAc. Cytokeratins, a large family of intermediate filaments present mostly in epithelial cells, are glycosylated by O-GlcNAc (Table 1). Initially, O-GlcNAc was demonstrated on cytokeratin 13 (King and Hounsell 1989). Later, cytokeratins 8 and 18 were also found to be multiply glycosylated (Chou et al. 1992; Chou and Omary 1993). As indicated above, the O-GlcNAcs on cytokeratins are substantially more dynamic than the polypeptides. Two-dimensional gel analyses show that the O-GlcNAc-modified cytokeratins elute with the major Coomassie-stained bands, while the phosphorylated cytokeratins do not appear to contain saccharide and account for only a small portion of the overall mass of proteins. The neurofilaments (NF-L, NF-M, and NF-H) from rat and mice spinal cords are also glycosylated (Dong et al 1993). Interestingly, most of the glycosylation sites are in the head domains of the filaments, in sites already shown to be functionally important by site-directed mutagenesis (Fig. 3) (Gill et al. 1990; Wong and Cle-

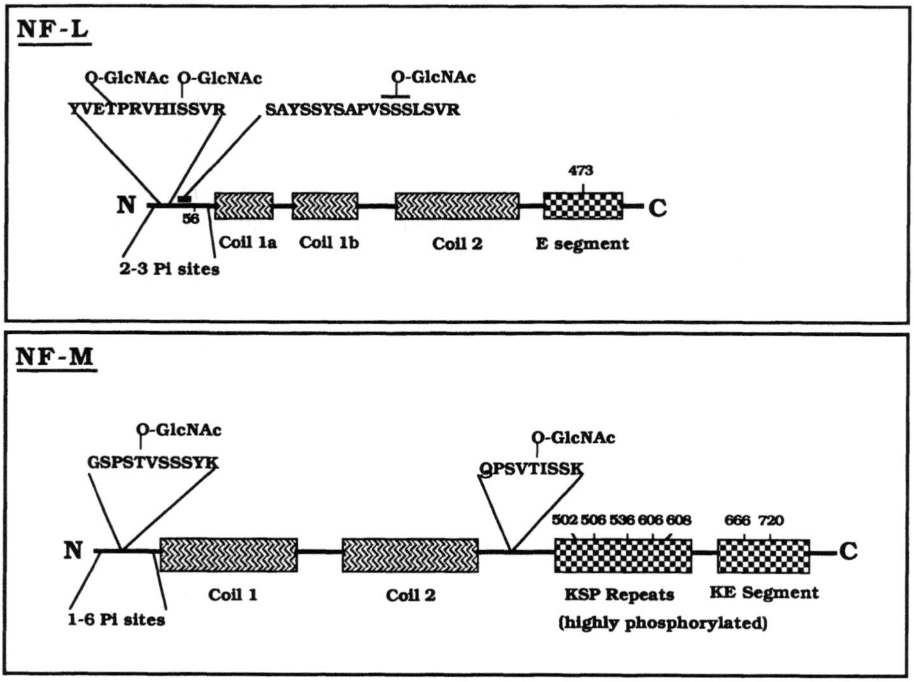

Fig. 3. Localization of O-GlcNAc on neurofilaments. Schematic diagrams of O-glycosylation and phosphorylation of NF-L (*top panel*) and NF-M (*bottom panel*). Identified O-GlcNAc sites and their surrounding tryptic sequences are shown. Known phosphorylation sites are *marked with numbers* corresponding to amino acid postion in rat NF-L and NF-M as follows: serine *56* (equivalent to serine 55 of mouse NF-L), serine *473* of NF-L, and serines *502, 506, 606, 608, 666* and *720* of NF-M. There are two to three phosphorylation sites on the head domain of NF-L and one to six sites on the head domain of NF-M. These are indicated as phosphorylation (*Pi*) sites

veland 1990). In fact, deletion of a 12-amino-acid sequence from the head domain of NF-L that happens to contain the two major glycosylation sites causes a dominant mutation that disrupts normal neurofilament assembly when the gene is transiently expressed in cells. Glycosylation site-specific mutagenesis of these sites is underway to determine more directly if the saccharide is involved in neurofilament assembly/disassembly processes. O-GlcNAc modification of lens α-crystallins (Roquemore et al. 1992), which are small heat shock proteins that form lens fibers, may also play a role in polymerization/depolymerization of these important structural components.

5 Viral Proteins

As for many of the cytoskeletal proteins, the O-GlcNAc moieties on the adenovirus fiber proteins are difficult to detect via galactosylation unless the fibers (trimers) are extensively denatured (Mullis et al. 1990), suggesting that the saccharide is "burried" when the protein trimerizes. The cytomegalovirus (CMV) basic phosphoprotein (BPP) is a major O-GlcNAc-bearing protein in CMV (Benko et al. 1988). While BPP's function is not well understood, it appears to play an important role in the initial assembly of the capsid proteins around the virus prior to its envelopment. Again, consistent with a possible role of O-GlcNAc in mediating polypeptide assembly reactions, other recently identified O-GlcNAc-bearing viral proteins include the NS26 protein of rotaviruses (Gonzalez and Burrone 1992) and the tegument protein of baculovirus (Whitford and Faulkner 1992). Insect intracellular proteins and baculovirus expressed proteins appear to be efficiently glycosylated by O-GlcNAc. Insect systems are particularly useful for the study of O-GlcNAc because of the low abundance of other glycoproteins with terminal GlcNAc residues. However, it is not yet known whether the same sites are modified in baculovirus expressed proteins that are utilized by mammalian cells in situ. The cell type, metabolic state, or cell-cycle stage dependence of the location of O-GlcNAc attachment sites is an important issue, particularly for the study of O-GlcNAc on low-abundance proteins, such as transcription factors or oncogenes, which requires overexpression of the proteins in artificial systems for biochemical analyses.

6 Enzymes of O-GlcNAc Addition and Removal

Using synthetic peptide substrates based upon known sites of O-GlcNAc addition (Table 2), a cytosolic O-GlcNAc transferase has been identified (Haltiwanger et al. 1990), and subsequently purified to homogeneity (Haltiwanger et al. 1992a). The transferase uses UDP-GlcNAc as its donor substrate (K_m~400 nM), has its active site in the cytosol based upon latency studies, and appears to be a trimer of ~340 kDa. SDS-PAGE and photoaffinity labeling studies, together with gel filtration and sedimentation analysis, suggest that the enzyme consists of two ~110-kDa catalytic subunits and a ~78-kDa putative regulatory subunit. The low turnover number of the purified enzyme suggests that the 78-kDa subunit may be negative regulator, but this remains to be tested. Preliminary studies early in the purification of the enzyme indicate that multiple forms of the transferase may exist, suggesting that, analogous to kinases, there may be a family of O-GlcNAc transferases. Low stringency screening using cDNAs from the rat liver enzyme will directly address this issue. The O-GlcNAc transferase has also been detected in yeast. Recently, it was shown that cell-free reticulocyte lysates contain enough endogenous UDP-GlcNAc and O-GlcNAc transferase activity to efficiently glycosylate in vitro translated p62 nucleoporin (Starr and Hanover 1990). Using this knowledge, a coupled cell-free transcription/ translation system, in conjunction with WGA-Sepharose chromatography, have been employed to detect and study the O-GlcNAc on rare transcription factors or oncogenes for which a cDNA is available. This approach provides a rapid screening method for any protein that has already had its cDNA cloned.

A cytosolic neutral N-acetylglucosaminidase has been purified from rat spleen (Dong et al. 1993 in prep.). The enzyme has selectivity toward O-GlcNAc peptides, and unlike lysosomal hexosaminidases, does not have activity against terminal GalNAc residues. The enzyme appears to be a dimer of a ~53-kDa and a ~55-kDa subunit, latency studies support its cytosolic localization, and it is potently inhibited by a number of compounds, including 1-amino-GlcNAc and 2-acetamido-deoxynorjirimycin. Such inhibitors should prove useful in evaluating the role of O-GlcNAc dynamics.

7 Conclusions

Although only recently discovered, we know that the posttranslational modification of intracellular proteins by O-GlcNAc is as abundant and apparently as dynamic as protein phosphorylation. Thus, O-GlcNAc is very distinct from known types of extracellular glycosylation. The presence of O-GlcNAc on so many proteins provides the potential of an entirely unexpected level of signal transduction or metabolic control. Even though direct data on function are currently lacking, the circumstantial evidence is rapidly mounting that O-GlcNAc plays a key role in the metabolism of eukaryotic cells. The tools for directly testing the specific role(s) of O-GlcNAc in transcriptional control, nuclear transport, cytoskeleton dynamics, and protein turnover are rapidly becoming available. Cloning of the O-GlcNAc transferase and the O-GlcNAcase will allow rapid progress with respect to general functions of this ubiquitous modification.

Acknowledgments. Original work is supported by NIH grants R01 HD13563 and R01 CA42486 to G.W.H., and W.G.K. was supported by the March of Dimes. D.L.-Y.D. was partially supported by a predoctoral Merck Foundation Fellowship. K.G. was a Johns Hopkins/Monsanto Postdoctoral Fellow.

References

Abeijon C, Hirschberg CB (1988) Intrinsic membrane glycoproteins with cytosol-oriented sugars in the endoplasmic reticulum. Proc Natl Acad Sci USA 85:1010–1014

Ambron RT, Protic J, Den H, Gabel CA (1989) Identification of protein-bound oligosaccharides on the surface of growth cones that bind muscle cells. J Neurobiol 20:549–568

Benko DM, Haltiwanger RS, Hart GW, Gibson W (1988) Virion basic phosphoprotein from human cytomegalovirus contains O-linked N-acetylglucosamine. Proc Natl Acad Sci USA 85:2573–2577

Chou C-F, Smith AJ, Omary MB (1992) Characterization and dynamics of O-linked glycosylation of human cytokeratin 8 and 18. J Biol Chem 267:3901–3906

Chou C-F, Omary MB (1993) Mitotic arrest-associated enhancement of O-linked glycosylation and phosphorylation of human keratins 8 and 18. J Biol Chem 268:4465–4472

Corden JL (1990) Tails of RNA polymerase II. TIBS 15:383–387

D'Onofrio M, Starr CM, Park MK, Holt GD, Haltiwanger RS, Hart GW, Hanover JA (1988) Partial cDNA sequence encoding a nuclear pore protein modified by O-linked N-acetylglusosamine. Proc Natl Acad Sci USA 85:9595–9599

Datta B, Ray MK, Chakrabarti D, Wylie DE, Gupta NK (1989) Glycosylation of eukaryotic peptide chain initiation factor 2 (eIF-2)-associated 67-kDa polypeptide (p^{67}) and its possible role in the inhibition of eIF-2 kinase-catalyzed phosphorylation of the eIF-2 α-subunit. J Biol Chem 264:20620–20624

Davis LI, Blobel G (1987) Nuclear pore complex contains a family of glycoproteins that includes p62: glycosylation through a previously unidentified cellular pathway. Proc Natl Acad Sci USA 84:7552–7556

Dong L-YD, Xu Z-S, Chevrier MR, Cotter RJ, Cleveland DW, Hart GW (1993) Glycosylation of mammalian neurofilaments. Localization of multiple O-linked N-acetylglucosamine moieties on neurofilament polypeptides L and M. J Biol Chem 268:

Finlay DR, Newmeyer DD, Price TM, Forbes DJ (1987) Inhibition of in vitro nuclear transport by a lectin that binds to nuclear pores. J Cell Biol 104:189–200

Finlay DR, Meier E, Bradley P, Horecka J, Forbes DJ (1991) A complex of nuclear pore proteins required for pore function. J Cell Biol 114:169–183

Forbes DJ (1992) Structure and function of the nuclear pore complex. Annu Rev Cell Biol 8:495–527

Gabel CA, Den H, Ambron RT (1989) Characterization of protein-linked glycoconjugates produced by identified neurons of *Aplysia californica.* J Neurobiol 20:530–548

Gill SR, Wong PC, Monteiro MJ, Cleveland DW (1990) Assembly properties of dominant and recessive mutations in the small mouse neurofilament (NF-L) subunit. J Cell Biol 111:2005–2019

Gonzalez SA, Burrone OR (1992) Rotavirus NS26 is modified by addition of single O-linked residues of N-acetylglucosamine. Virology 182:8–16

Hagmann J, Grob M, Burger MM (1992) The cytoskeletal protein talin is O-glycosylated. J Biol Chem 267:14424–14428

Haltiwanger RS, Holt GD, Hart GW (1990) Enzymatic addition of O-GlcNAc to nuclear and cytoplasmic proteins: identification of an uridine diphospho-N-acetylglucosamine: peptide B-N-acetylglucosamyltransferase. J Biol Chem 265:1–6

Haltiwanger RS, Blomberg MA, Hart GW (1992a) Glycosylation of nuclear and cytoplasmic proteins. Purification and characterization of a uridine diphospho-N-acetylglucosamine: polypeptide β-N-acetylglucosaminyltransferase. J Biol Chem 267:9005–9013

Haltiwanger RS, Kelly WG, Roquemore EP, Blomberg MA, Dong L-YD, Kreppel L, Chou T-Y, Hart GW (1992b) Glycosylation of nuclear and cytoplasmic proteins is ubiquitous and dynamic. Biochem Soc Trans 20:264–269

Hanover JA, Cohen CK, Willingham MC, Park MK (1987) O-linked N-acetylglucosamine is attached to proteins of the nuclear pore. Evidence for cytoplasmic and nucleoplasmic glycoproteins. J Biol Chem 262:9887–9894

Hanover JA (1992) The nuclear pore: at the crossroads. FASEB J 6:2288–2295

Hart GW, Haltiwanger RS, Holt GD, Kelly WG (1989) Glycosylation in the nucleus and cytoplasm. Annu Rev Biochem 58:841–874

Hirai S, Kawasaki H, Yaniv M, Suzuki K (1991) Degradation of transcription factors, c-Jun and c-Fos, by calpain. FEBS Lett 287:57–61

Holt GD, Hart GW (1986) The subcellular distribution of terminal N-acetylglucosamine moieties. Localization of a novel protein-saccharide linkage, O-linked GlcNAc. J Biol Chem 261:8049–8057

Holt GD, Snow CM, Senior A, Haltiwanger RS, Gerace L, Hart GW (1987) Nuclear pore complex glycoproteins contain cytoplasmically disposed O-linked N-acetylglucosamine. J Cell Biol 104:1157–1164

Jackson SP, Tjian R (1988) O-Glycosylation of eukaryotic transcription factors: implications for mechanisms of transcriptional regulation. Cell 55:125–133

Jackson SP, Tjian R (1989) Purification and analysis of RNA polymerase II transcription factors by using wheat germ agglutinin affinity chromatography. Proc Natl Acad Sci USA 86:1781–1785

Kearse KP, Hart GW (1991a) Lymphocyte activation induces rapid changes in nuclear and cytoplasmic glycoproteins. Proc Natl Acad Sci USA 88:1701–1705

Kearse KP, Hart GW (1991b) Topology of O-linked N-acetylglucosamine in murine lymphocytes. Arch Biochem Biophys 290:543–548

Kelly WG, Hart GW (1989) Glycosylation of chromosomal proteins: localization of O-linked N-acetylglucosamine in *Drosophila* chromatin. Cell 57:243–251

Kelly WG, Dahmus ME, Hart GW (1993) RNA polymerase II is a glycoprotein: modification of the C-terminal domain by O-GlcNAc. J Biol Chem 268 (in press)

King IA, Hounsell EF (1989) Cytokeratin 13 contains O-glycosidically linked N-acetylglucosamine residues. J Biol Chem 264:14022–14028

Lichtsteiner S, Schibler U (1989) A glycosylated liver-specific transcription factor stimulates transcription of the albumin gene. Cell 57:1179–1187

Luthi T, Haltiwanger RS, Greengard P, Bahler M (1991) Synapsins contain O-linked N-acetylglucosamine. J Neurochem 56:1493–1498

McClain DA, Paterson AJ, Roos MD, Wei X, Kudlow JE (1992) Glucose and glucosamine regulate growth factor gene expression in vascular smooth muscle cells. Proc Natl Acad Sci USA 89:8150–8154

Meek DW, Street AJ (1992) Nuclear protein phosphorylation and growth control. Biochem J 287:1–15

Meikrantz W, Smith DM, Sladicka MM, Schlegel RA (1991) Nuclear localization of an O-glycosylated protein phosphotyrosine phosphatase from human cells. J Cell Sci 98:303–307

Mullis KG, Haltiwanger RS, Hart GW, Marchase RB, Engler JA (1990) Relative accessibility of N-acetylglucosamine in trimers of the adenovirus types 2 and 5 fiber proteins. J Virol 64:5317–5323

Newmeyer DD, Forbes DJ (1988) Nuclear import can be separated into distinct steps in vitro: nuclear pore binding and translocation. Cell 52:641–653

Nyame K, Cummings RD, Damian RT (1987) Schistosoma mansoni synthesizes glycoproteins containing terminal O-linked N-acetylglucosamine residues. J Biol Chem 262:7990–7995

Park MK, D'Onofrio M, Willingham MC, Hanover JA (1987) A monoclonal antibody against a family of nuclear pore proteins (nucleoporins): O-linked N-acetylglucosamine is part of the immunodeterminant. Proc Natl Acad Sci USA 84:6462–6466

Privalsky ML (1990) A subpopulation of the avian erythroblastosis virus v-erbA protein, a member of the nuclear hormone receptor family, is glycosylated. J Virol 64:463–466

Ralph RK, Darkin-Rattray S, Schofield P (1990) Growth-related protein kinases. BioEssays 12:121–124

Reason AJ, Morris HR, Panico M, Marais R, Treisman RH, Haltiwanger RS, Hart GW, Kelly WG, Dell A (1992) Localization of O-GlcNAc modification on the serum response transcription factor. J Biol Chem 267:16911–16921

Roach PJ (1991) Multisite and hierarchal protein phosphorylation. J Biol Chem 266:14139–14142

Rogers S, Wells R, Rechsteiner M (1986) Amino acid sequences common to rapidly degraded proteins: the PEST hypothesis. Science 234:364–368

Roquemore EP, Dell A, Morris HR, Panico M, Reason AJ, Savoy L-A, Wistow GJ, Zigler JS Jr, Earles BJ, Hart GW (1992) Vertebrate lens α-crystallins are modified by O-linked N-acetylglucosamine. J Biol Chem 267:555–563

Schindler M, Hogan M, Miller M, DeGaetano D (1987) A nuclear specific glycoprotein representative of an unique pattern of glycosylation. J Biol Chem 262:1254–1260.

Snow CM, Senior A, Gerace L (1987) Monoclonal antibodies identify a group of nuclear pore complex glycoproteins. J Cell Biol 104:1143–1156

Starr CM, D'Onofrio M, Park MK, Hanover JA (1990) Primary sequence and heterologous expression of nuclear pore glycoprotein p62. J Cell Biol 110:1861–1871

Starr CM, Hanover JA (1990) Glycosylation of nuclear pore protein p62. Reticulocyte lysate catalyzes O-linked N-acetylglucosamine addition in vitro. J Biol Chem 265:6868–6873

Taylor SS, Adams JA (1992) Protein kinases: coming of age. Curr Opinion Struct Biol 2:743–748

Torres C-R, Hart GW (1984) Topography and polypeptide distribution of terminal N-acetylglucosamine residues on the surfaces of intact lymphocytes. J Biol Chem 259,5:3308–3317

Whiteheart SW, Passaniti A, Reichner JS, Holt GD, Haltiwanger RS, Hart GW (1989) Glycosyltransferase probes. Methods Enzymol 179:82–95

Whitford M, Faulkner P (1992) A structural polypeptide of the baculovirus *Autographa californica* nuclear polyhedrosis virus contains O-linked N-acetylglucosamine. J Virol 66:3324–3329

Wong PC, Cleveland DW (1990) Characterization of dominant and recessive assembly-defective mutations in mouse neurofilament NF-M. J Cell Biol 111:1987–2003

Novel N-Glycosylation in Eukaryotes: Laminin Contains the Linkage Unit β-Glucosylasparagine

R. Schreiner[1], Eva Schnabel[2], and F. Wieland[3]

1 Introduction

Glycoproteins represent the major proportion of mammalian proteins. Whereas a variety of O-glycosyl-linkage units exist between carbohydrate and the protein in eukaryotes, N-glycosylation seemed to be restricted to one type of linkage unit: N-acetylglucosamine in β linkage to the amido-nitrogen of asparagine (β-GlcNAcAsn). Although an N-glycosyl linkage unit made of glucose and Asn with the glucose in α-anomeric form (α-GlcAsn) has been suggested to occur in eukaryotes (Shibata et al. 1988), this type of linkage has not been unequivocally proven to exist. On the other hand, the extreme halophilic archebacteriae *H. halobium* and *H. volcanii* were shown to contain a linkage unit β-GlcAsn, as established by its isolation from the cell surface glycoproteins of these prokaryotic organisms (Wieland et al. 1983; Lechner and Wieland 1990) and by subsequent chemical characterization.

In our attempt to analyze mammalian proteins for the existence of α- or β-GlcAsn linkages, we chemically synthesized α- and β-glucosylasparagine, coupled these compounds to a carrier protein, and obtained polyclonal antibodies against both anomers. These antibodies show no crossreactivity with the unit β-GlcNAcAsn, and they are specific for their α- or β-anomeric antigen, respectively.

We report here that the N-glycosyl-protein linkage unit β-GlcAsn indeed exists in mammals, specifically in basement membranes, and establish this immunological finding by isolation of the compound from the basement membrane component laminin and by its chemical characterization. Thus, β-GlcAsn is a second type of N-glycosyl linkage unit in eukaryotes (Schreiner et al. 1994).

2 Characterization of Antisera Against α-GlcAsn and β-GlcAsn

α- and β-GlcAsn were synthesized according to Waldmann and Kunz (1983) and were blocked at their carboxy termini by amidation. These compounds were linked to presuccinylated bovine serum albumin via their N-termini (Samokhin and Filimonov 1985). About 40 residues and less were bound per carrier molecule. The conjugates were used to raise rabbit antibodies. Figure 1 shows their specificity as characterized

[1] Baylor Research Institute, 3812 Elm Street, Dallas, Texas 75226, USA.
[2] Anton Pilgrim Weg 8, D-71701 Schwieberdingen, FRG.
[3] Institut für Biochemie I der Universität Heidelberg, Im Neuenheimer Feld 328, D-69120 Heidelberg FRG.

44. Colloquium Mosbach 1993
Glyco- and Cellbiology

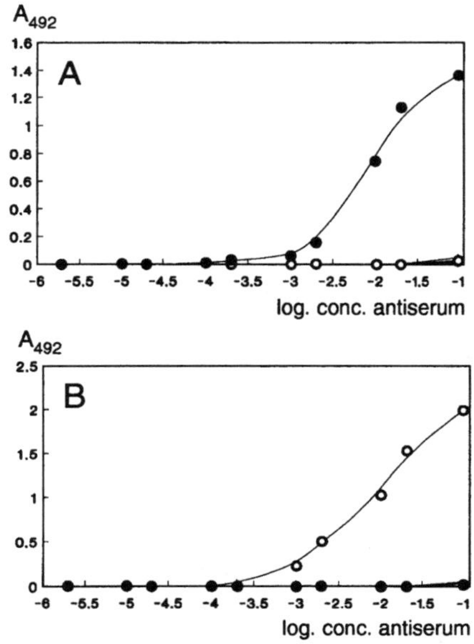

Fig. 1 A, B. Characterization of the anti-α-GlcAsn and the anti-β-GlcAsn antibodies. **A** Affinity purified anti-α-AsnGlc antibodies were tested by ELISA against their antigen α-GlcAsn-succ-BSA (l) and against β-GlcAsn-succ-BSA (m). In addition, the following antigens were analyzed: β-GalAsn-succ-BSA, β-RhamAsn-succ-BSA, β-GlcNAcAsn-succ-BSA, and β-GalNAcAsn-succ-BSA. **B** The affinity purified anti-βGlcAsn antibodies were tested against their antigen β-GlcAsn-succ-BSA(m) and against α-GlcAsn-succ-BSA (l), and against the additional antigens listed under **A**

by enzyme linked immuno assay (ELISA). The anti-α-GlcAsn antiserum was highly specific for the α-GlcAsn-conjugate and did not cross-react with the β-conjugate (Fig. 1 A). Likewise, antibodies obtained after injection of the β-conjugate showed high specificity for the β-anomer. (Fig. 1 B). The high specificity of both antisera was also shown by their lack of reactivity with the following N-glycosyl-asparagine derivatives, all bound to succ-BSA: galactosyl β(1–N) asparagine (GalAsn), L-rhamnosyl β(1–N) asparagine (RhamAsn), N-acetyl glucosaminyl β(1–N) asparagine (GlcNAcAsn), and N-acetyl-galactosaminyl β(1–N) asparagine (GalNAcAsn). These antigens gave only background readings in the ELISA in the range of concentrations of the antibodies used, and therefore are not included in Fig. 2.

The linkage unit β-GlcAsn had originally been found in the cell surface glycoprotein of the archaebacterium *H. halobium* (Wieland et al. 1983). In this protein, the N-linked glucose bears an additional carbohydrate structure linked to position 4 of the glucose residue. In order to further characterize the antisera against β-GlcAsn, these additional sugars must be removed from the archebacterial glycoprotein, thereby "unmasking" the antigenic epitope. This removal is easily achieved under mild conditions by solvolysis in anhydrous HF, a treatment which yields an intact protein core that still contains those sugar monomers that are linked in N-glycosyl units. Reaction of the antibodies with this antigen was analyzed by Western blotting of untreated and HF-solvolyzed halobacterial cell surface glycoprotein (Fig. 2). After solvolysis with HF, the anti-β-GlcAsn antiserum bound to the cell surface glycoprotein (Fig. 2, lane 8). In contrast, anti-β-GlcAsn antibodies did not recognize any antigenic determinants in proteins from halobacteria as long as these were not treated with HF (Fig. 2, lane 7). As an additional control of the specificity of the antibodies, nonenzymatically glucosylated BSA and fetuin were analyzed before and after treatment with HF. Fetuin that contains the common linkage β-GlcNAcAsn unit of N-glycosylated proteins did

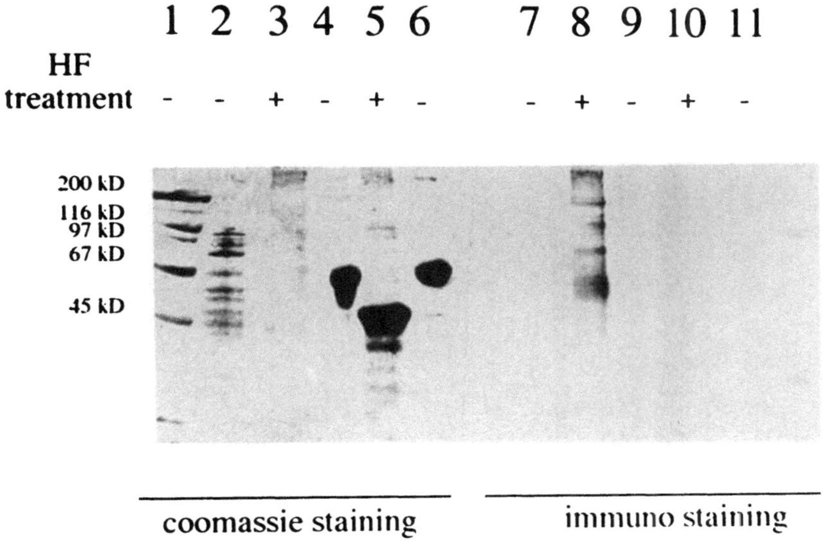

Fig. 2. Specificity of the anti-β-GlcAsn antibodies assessed by Western blot analysis. Various glycoproteins were separated by SDS gel electrophoresis and subsequently electrotransferred to Immobilon sheets. Proteins were visualized by staining with Coomassie Blue (*lanes 1–6*) or by immunoreaction (*lanes 7–11*) with the anti-β-GlcAsn antibodies. *Lane 1* Molecular weight standard, *lanes 2 + 7* cell envelope proteins of *Halobacterium halobium*, 30 μg each; *lane 3* 30 μg, and *lange 8* 5 μg HF-treated cell envelope proteins of *Halobacterium halobium*; *lanes 4 + 9* fetuin, 30 μg each; *lanes 5 + 10* HF-treated fetuin, 40 μg each; *lanes 6 + 11* nonenzymatically glucosylated BSA, 30 μg each

not react with the anti-β-GlcAsn antiserum, either before or after solvolysis (Fig. 2, lanes 9 and 10). The same result was obtained with the nonenzymatically glucosylated BSA (Fig. 2, lane 11). A similar blot (like Fig. 2, lanes 7–11) incubated with anti-α-GlcAsn antiserum was completely blank and therefore is not shown here.

The HF-solvolyzed cell surface glycoprotein from halobacteria or their HF-solvolyzed flagellae (Wieland et al. 1985) served as a matrix for immuno affinity purification of the anti-β-GlcAsn antibodies. The resulting immunoaffinity purified antibodies were used for the studies described below. Anti-α-antibodies were purified by neutralizing with succ-BSA and subsequent immunoadsorption to α-GlcAsn-succ-BSA.

3 Immunological Characterization of Kidney Glomeruli Extracts

Shibata et al. (1988) have described a nephritogenic glycopeptide from a rat kidney glomerulus fraction which contained predominantly glucose. They proposed the carbohydrate moiety of this glycopeptide to be attached to an Asn residue in α-anomeric linkage, although this has only been concluded from NMR spectroscopy of an extended tryptic glycopeptide, without isolation and characterization of the linkage unit.

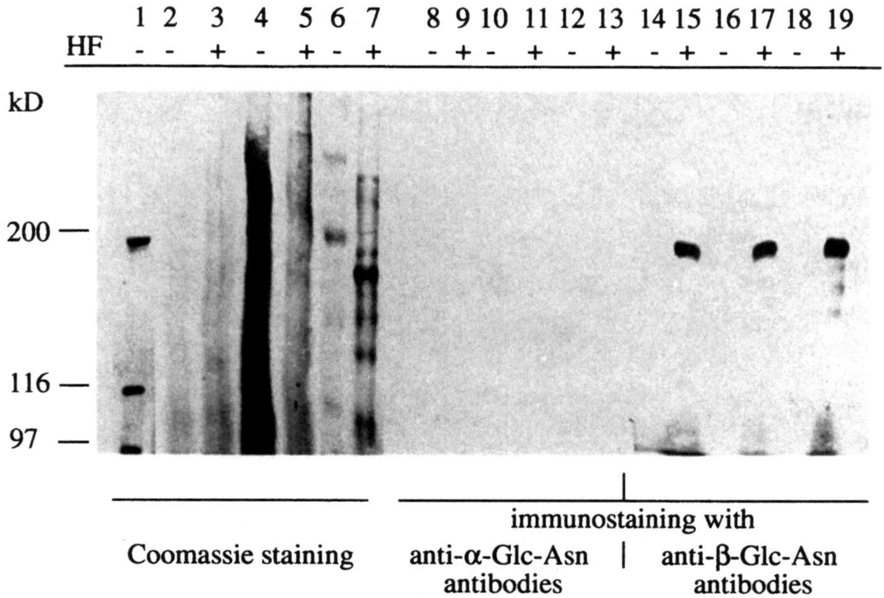

Fig. 3. Immunological analysis by Western blotting with anti-α-GlcAsn and anti-β-GlcAsn antibodies of basal membrane proteins. Protein extracts of glomeruli and isolated laminin were separated by SDS gel electrophoresis and subsequently electrotransferred to Immobilon sheets. Thereafter the proteins were visualized by staining either with Coomassie (*lanes 1–7*) or by immunoreaction either with anti-α-GlcAsn antibodies (*lanes 8–13*) or with anti-β-GlcAsn antibodies (*lanes 14–19*). *Lane 1* Molecular weight standard; *lanes 2, 8, and 14* protein extract from rat kidney glomeruli; *lanes 3, 9 and 15* HF-treated protein extract from rat kidney glomeruli; *lanes 4, 10 and 16* protein extract from porcine kidney glomeruli; *lanes 5, 11 and 17* HF-treated protein extract from porcine kidney glomeruli; *lanes 6, 12 and 18* laminin; *lanes 7, 13 and 19* HF-treated laminin

Therefore we have used the anti-α- and anti-β-GlcAsn antisera to screen kidney glomeruli by Western blotting for the occurence of these epitopes. The result is shown in Fig. 3. Extracts of glomeruli from rat and pig do not react at all with the anti-α-GlcAsn antiserum or the anti-β-GlcAsn-antibodies (Fig. 3, lanes 8 + 10 and 14 + 16). However, after solvolysis with HF, these extracts contained proteins that were stained with the anti-β-GlcAsn antiserum. In both fractions, distinct staining was observed in a molecular weight range of close to 200 kDa (Fig. 3, lanes 15 and 17). As there is only a small number of proteins in this molecular weight range, we have immunologically analyzed the following isolated basement membrane components of similar size, using the β-GlcAsn antibodies: heparansulfate proteoglycan, collagen, fibronectin, and laminin, each before and after HF treatment. Tested on the purified proteins, the anti-β-GlcAsn antibodies gave a clearcut positive response only with HF-treated laminin (Fig. 3, lane 19). Again, the anti-α-antibodies did not react (Fig. 3, lane 12 and 13). These immunological results encouraged us to attempt an isolation from laminin of the linkage unit β-GlcAsn and its chemical characterization.

4 Isolation and Characterization of a Linkage Unit β-GlcAsn from Laminin

Purified mouse laminin (from an EHS tumor) was digested with trypsin and subsequntly loaded on a Sep-Pak cartridge. The cartridge was developed with water and subsequently with various concentrations of acetonitrile in water. These fractions were divided, and one half of each was solvolyzed with HF. Thereafter, the samples were digested with pronase E and passed through cation exchange resin under conditions known to elute synthetic β-GlcAsn.

The resulting eluates were converted to their orthophthaldialdehyde (OPA) derivatives and analyzed by HPLC. To this end, OPA derivatives of various synthetic N-glycosyl linkage units were prepared as standard compounds. The resulting elution profile is shown in Fig. 4 B. β-GlcAsn (peak 3) is clearly separated from the other N-glycosol-linkage units, e.g., from α-GlcAsn (peak 1), and from β-GlcNAc-Asn (peak 5). In panel A, separation of an amino acid standard mixture is shown to demonstrate

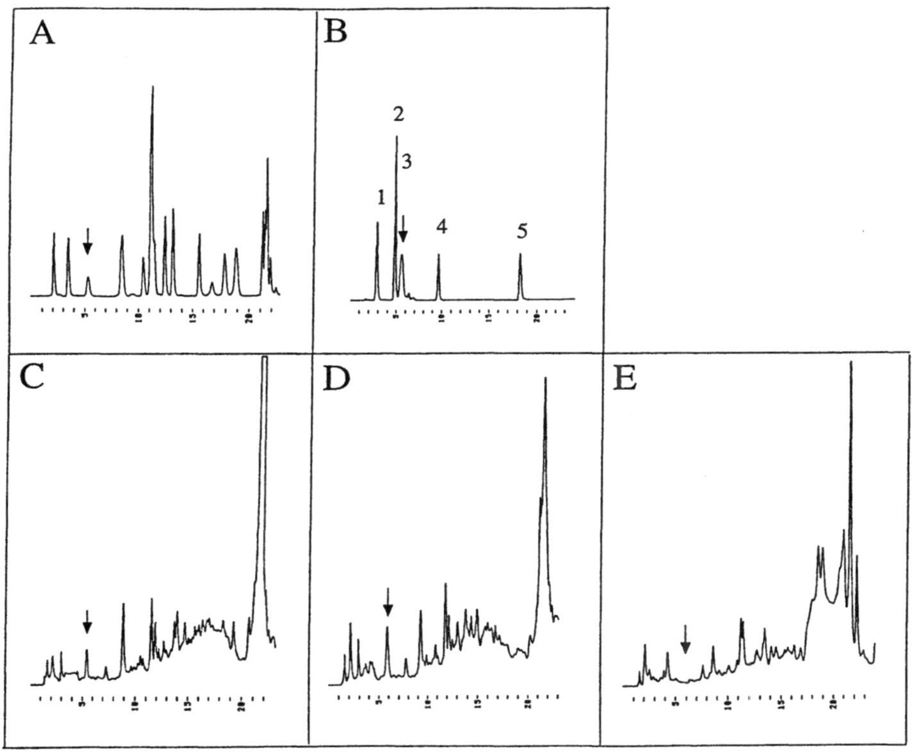

Fig. 4 A–E. Analysis of laminin derived peptiden 1 by comparison with authentic β-GlcAsn-standard. Sample and standard were reacted with ortho-phthaldialdehyde before separation by Reversed-Phase HPLC (λ = 455 nm). **A** Mixture of an amino acid standard solution and synthetic β-GlcAsn. **B** Mixture of glycosyl-asparagine derivatives: α-GlcAsn (*1*), β-GalAsn (*2*), β-GlcAsn (*3*), β-RhamAsn (*4*), and β-GlcNAcAsn (*5*). **C** HF-treated and pronase E-derived peptide fraction as in C, but without HF-treatment from laminin. **D** Mixture of the fraction shown in **C** and β-GlcAsn (ratio 1:1). **E** Fraction as in C, but without HF-treatment

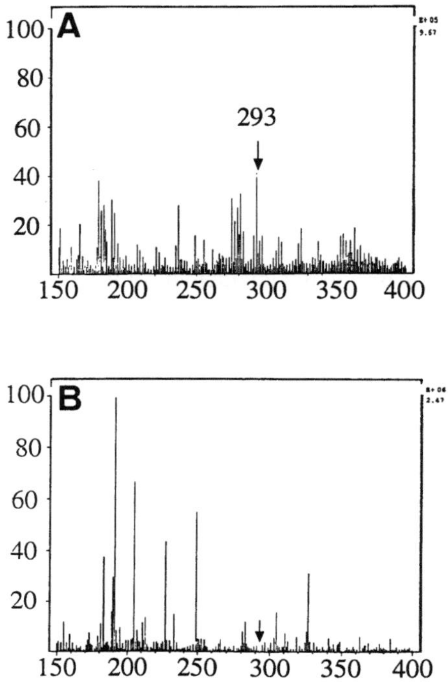

Fig. 5 A, B. FAB-MS of the pronase E-derived peptide fraction from laminin after (**A**) and before HF treatment (**B**). Tryptic peptides were digested with pronase and purified on a Dowex ion exchanger, and thereafter treated with HF, and analyzed by FAB-MS run in the negative mode. **A** HF-treated sample. **B** Untreated control sample

that no natural amino acid-derivatives elute in the region of β-GlcAsn (see arrow in Fig. 4 A). Separation of the OPA-derivatized laminin derived peptide fraction is depicted in Fig. 4 C and E, respectively. A peak is observed at the retention time of authentic β-GlcAsn (arrow in panel C) exclusively in the sample that had been treated with HF before digestion with pronase and cation exchange chromatography [fraction I(+)]. Fraction I(–) is devoid of this material. Addition of synthetic β-GlcAsn standard to fraction I(+) and subsequent HPLC yielded the profile shown in Fig. 4 D. No additional peak appeared, but the area of the peak in question is relatively enlarged and it shows no asymmetry. This strongly supports the idea that the compound β-GlcAsn actually is contained in fraction I(+). In order to obtain additional and independent evidence for the occurence of this compound in laminin, trypsin-derived laminin peptides were partially separated by HPLC, and the resulting fractions were solvolyzed by HF and analyzed by ELISA for their content of β-GlcAsn. Two positive fractions were obtained. The major fraction was digested with pronase E, and the resulting aminoacylsaccharide purified by cation exchange chromatography as described above. The eluate was dried and subjected to FAB-mass spectroscopy in the negative mode. The resulting mass spectrum is shown in Fig. 5. A mass peak appears at 293.3 in the HF treated sample, corresponding to M-1 of asparagine connected to a neutral hexose (Fig. 5 A). In contrast, the untreated sample showed no such signal (Fig. 5 B).

Finally, the retention time of synthetic β-GlcAsn by GLC-MS on a 30-m capillary column was determined after methylation and pertrifluoroacetylation (Fig. 6 A). The

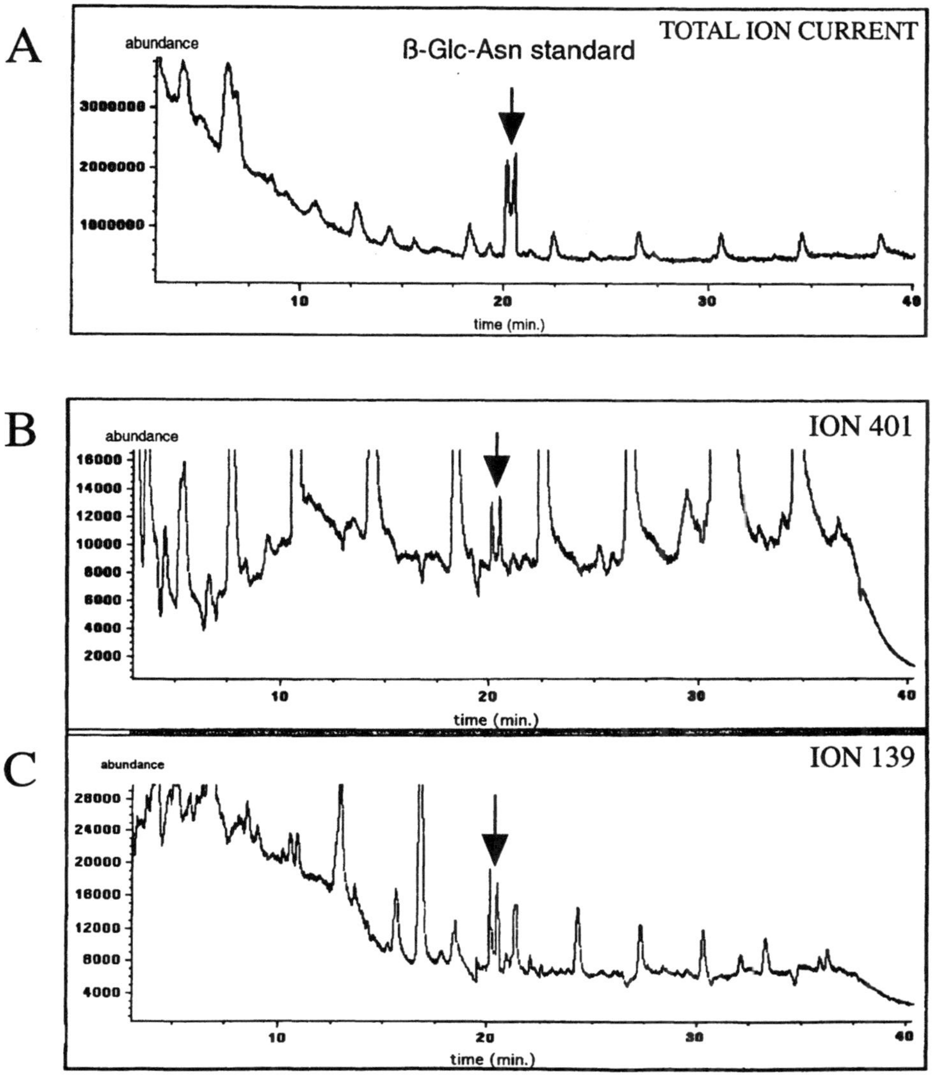

Fig. 6 A–C. GLC-MS of β-GlcAsn (**A**) and the pronase derived peptide fraction from laminin after HF treatment (**B** and **C**). **A** Total ion current of a GLC run of 20 μg synthetic standard β-GlcAsn monitored by a mass selective detector. **B, C** Single ion current after GLC run of HF treated, pronase treated fraction 4, scheme 1, at single ions set to a mass 401, and 139, respectively. GlC was performed on a Durabond 1701 capillary column in a Hewlett Packard combined gas liquid chromatograph-mass spectrometer

HF-treated sample from laminin (as used for FAB mass spectroscopy) was subjected to the same analysis in the single ion monitoring mode at the ions 401 (Fig. 6 B) and 139 (Fig. 6 C) characteristic for β-GlcAsn. In each case, the double peak characteristic for standard β-GlcAsn was observed at identical retention times (standard: 20.13 and 20.52 min, ion 401: 20.13 and 20.51 min, ion 139: 20.13 and 20.51 min).

5 Chemical O-Deglycosylation of Fixed Tissue

An epitope that is masked by carbohydrates obviously cannot be analyzed by immunofluorescence microscopy of tissue slices. Therefore we have worked out a method that allows chemical O-deglycosylation of fixed tissue blocks. The method takes advantage of the mild solvolysis of O-glycosidic bonds by HF, as explained above. To minimize the incubation times necessary for complete O-deglycosylation, we have quantitated the carbohydrate content of tissue blocks after deglycosylation for various times. The result is shown in Fig. 7. Maximal release of carbohydrate is observed after incubation of fixed tissue cubes (size about 2 mm^3) at 0 °C in HF/ethanol (1/1, vol/vol) for 3 to 4 h.

Therefore, for O-deglycosylation, tissue of similar size was routinely incubated for 4 h under the above-mentioned conditions.

The extent of damage to the cellular structures caused by this treatment was assessed by electronmicroscopy. Figure 8 A shows an area of an HF-treated section at low magnification stained with lead citrate and uranyl acetate. A peritubular capillary (CP) is depicted located between a distal tubulus (DT) and a proximal tubulus (PT).

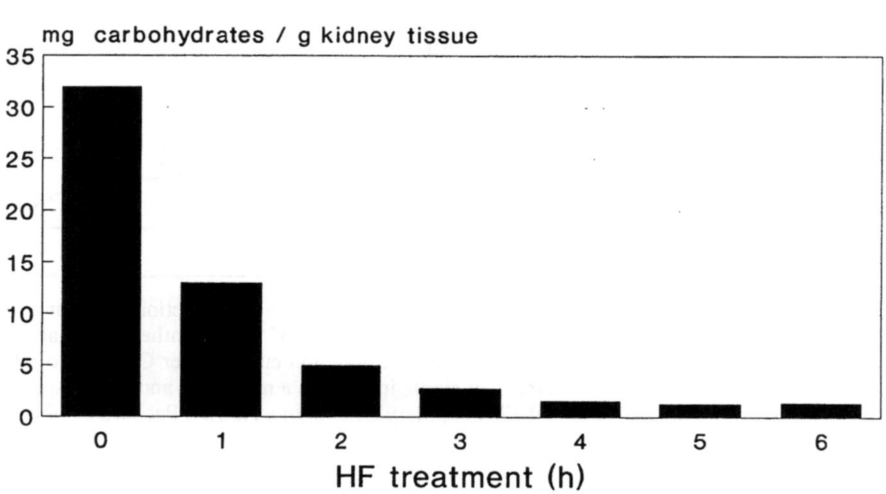

Fig. 7. Quantification of the carbohydrate content of tissue blocks after HF-treatment. Fixed tissue blocks of about 2 mm side length from rat kidney were treated with HF/ethanol for various times. After rehydration the tissue was homogenized and aliquots taken for the colorimetric quantification of carbohydrate with a modified anthrone method as described in Materials

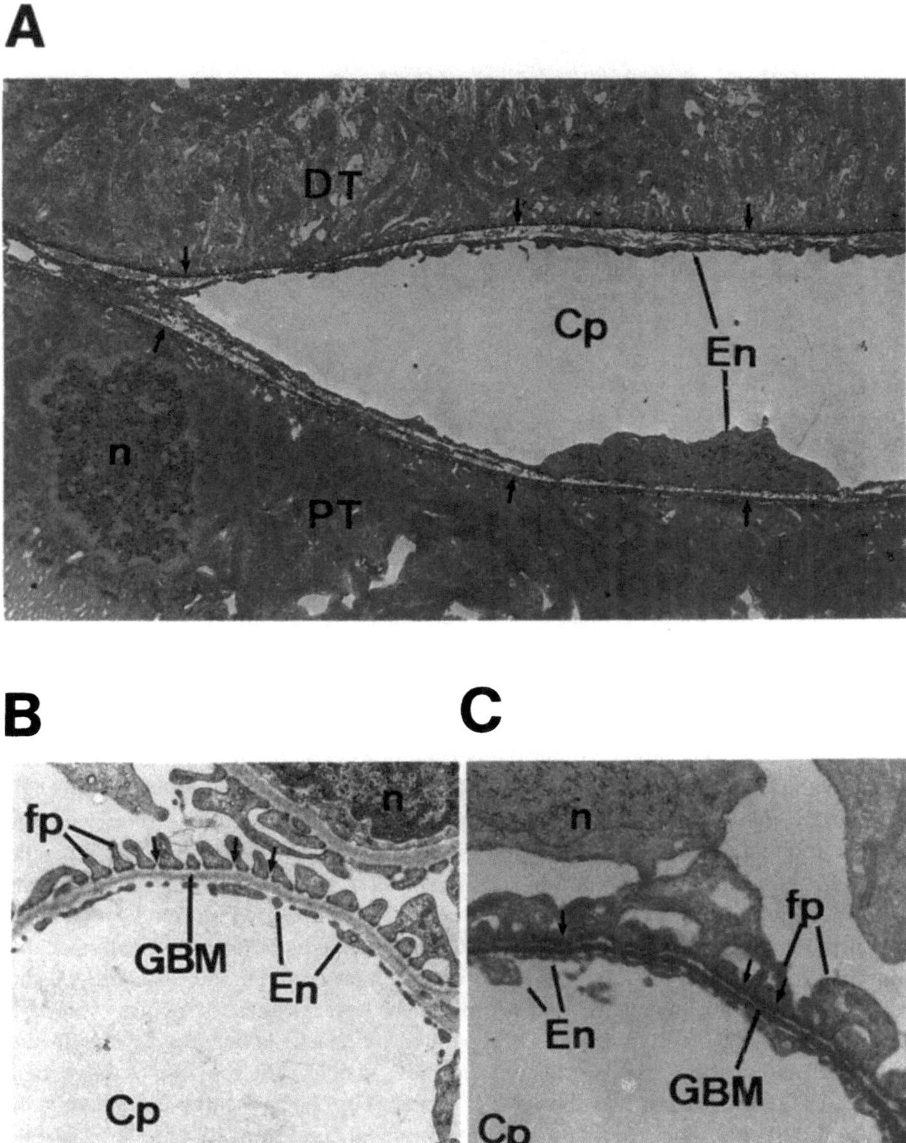

Fig. 8 A–C. Electromicroscopy of fixed rat kidney tissue after HF-treatment. Fixed tissue blocks from rat kidney were sovolyzed with HF/ethanol for 4 h and processed for electromicroscopy as described in Materials. **A** HF-treated section, magnification 5 400 fold. **B** Untreated section, 18 700 fold. **C** HF-treated section, 18 700 fold. Abbreviations: *Cp* peritubular capillary; *DT* distal tubulus; *PT* proximal tubulus; *En* fenestrated endothelium; *GBM* glomerular basement membrane; *fp* foot protrusions; *n* nucleus

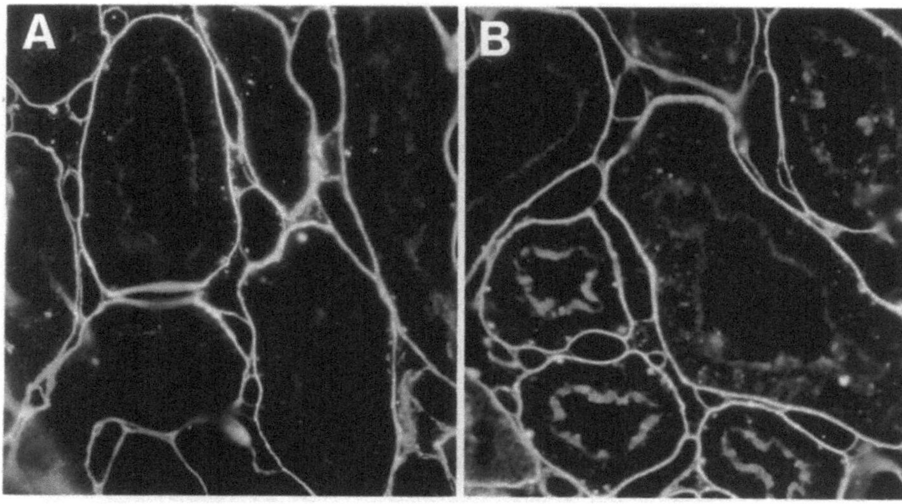

Fig. 9 A, B. Immunofluorescence microscopy with anti laminin (protein) antibodies of semithin sections from rat kidney before and after treatment with HF. Semithin frozen sections were taken and processed as described in Materials. First antibody: anti-laminin (protein) antibodies, second antibody: fluorescein-conjugated sheep anti rabbit IgG $F(ab)_2$. **A** Untreated tissue. **B** HF-treated tissue

The basement membrane of the tubular epithelium (arrows) appears unfragmented and regular. Figure 8B shows a control electronmicrograph of an untreated kidney tissue sample at higher magnification. A similar area at the same magnification after HF treatment is shown in Fig. 9 C. Although some alterations can be observed, the overall architecture of the HF-treated sample is still visible. Specifically, the glomerular basement membrane appears again as a regular and continuous linear structure. Due to the loss of carbohydrate, the overall staining with uranyl acetate and lead citrate is less intensive in the HF-treated sample. This holds specifically for the staining of the nucleus. Some influence can be observed on the structure of the podocytes, (Po) with their foot protrusions. In the control, these structures protrude from the basement membrane and are separated by spaces about 2 nm wide. In the HF-treated sample the foot protrusions are still visible, but the spaces have disappeared. This may be due to the lack of negatively charged carbohydrate (glycocalix) in the treated samples, that under physiological conditions by charge repulsion may maintain this spatial separation.

Thus, a fixed tissue block can be deglycosylated without severe destruction of its architecture, and therefore we were interested to analyze the immunological integrity of the deglycosylated samples. This was achieved by immunofluorescence microscopy on "semithin" (0.5–1 mm) kryosections.

In Fig. 9, the immunofluorescence of such slices is shown after incubation with an anti-laminin-antibody directed against laminin protein epitopes before (Fig. 9 A) and after (Fig. 10 B) treatment with HF. The untreated sample (Fig. 9 A) shows an extended linear fluorescence. The area depicts cross sections through proximal and distal

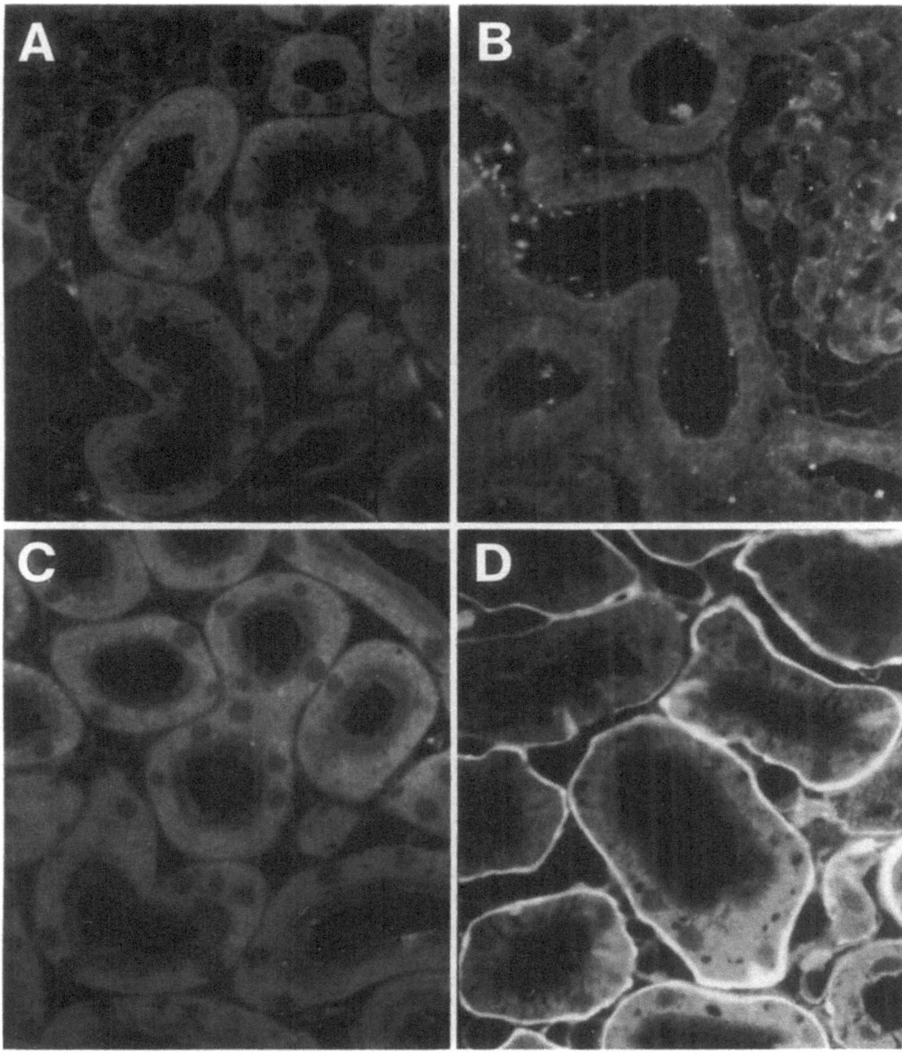

Fig. 10 A–D. The epitope β-GlcAsn can be visualized in kidney tissue after HF-treatment. Immunofluorescence microscopy was performed on frozen semithin sections from fixed kidney tissue and processed for immunofluorescence microscopy as described in Materials. **A** Untreated (control) tissue incubated with anti-α-GlcAsn-antibodies. **B** HF-treated tissue incubated with anti α-GlcAsn antibodies. **C** Untreated (control) tissue incubated with anti-β-GlcAsn antibodies. **D** HF-treated tissue incubated with anti-β-GlcAsn antibodies

tubuli. In Fig. 9 B, the HF-treated sample, no differences can be observed. Thus, laminin epitopes in the basement membrane are intact after HF treatment.

6 Localization of the Epitope β-GlcAsn to the Glomerular Basement Membrane

With these results and our finding in mind that anti β-GlcAsn antibodies bind to isolated laminin åfter HF solvolysis, we analyzed the localization of the novel type of N-glycosyl linkage by immunofluorescence of deglycosylated semithin kidney slices by reaction with the anti-β-GlcAsn antibodies, as shown in Fig. 10 C and D. As a control, incubations were performed with the anti-α-GlcAsn-antibodies (Fig. 10 A, B).

Exclusively in the HF-treated sample that was incubated with the anti-β-GlcAsn antibodies was an extended linear fluorescence signal observed (Fig. 10 D). The untreated sample showed only background staining (Fig. 10 C) (note that exposure for the photograph in Fig. 10 D was 3 s and in Fig. 10 A, B and C was 120 s). Likewise no signal was observed with the anti-α-GlcAsn-antibodies, either after of before HF treatment (Fig. 10 B, A).

By a combination of immunological and chemical analyses we established that laminin from kidney glomerular basement membranes contains the carbohydrate to protein linkage unit β-GlcAsn. A linkage unit α-GlcAsn that has been proposed in the literature could not be detected with our methods.

The antibodies used in this study were characterized:

1. by their reactivity against their chemically synthesized antigens;
2. by their lack of reactivity against the synthesized anomeric isoforms of their antigens;
3. by their lack of reactivity with protein-conjugated synthetic β-GlcNAcAsn as well as natural β-GlcNAcAsn as prepared by solvolysis with HF of conventional N-linked glycoproteins;
4. by their lack of reactivity against natural glycoprotein known to contain complex carbohydrates connected to protein by β-GlcAsn linkage, and their reactivity against these glycoproteins after solvolysis with HF, which is known to leave the N-linked monohexosyl units connected to the protein; and
5. by their lack of reactivity with nonenzymatically glucosylated proteins.

These various lines of immunological characterization almost exclude any unspecific reaction of the antibodies. Specifically, crossreactions would have been detected with the glycoproteins before solvolysis with HF.

In case of the cell surface glycoprotein of *Halobacterium halobium*, the existence of carbohydrates linked to the glucose residue of β-GlcAsn was established, and therefore the need for HF solvolysis to unmask the linkage unit was expected. Our finding that laminin also had to be treated with HF before its reaction with the anti-β-GlcAsn antibodies indicates that in laminin the glucose involved in this linkage carries additional sugars as well. These sugars have not been identified to date. Alternatively the epitope may be "masked" by neighboring oligosaccharides.

The mono-hexosyl-amino acid GlcAsn is a very small epitope, and the strong immunoresponse against this antigen may seem remarkable. At first glance, the exclusive antibody reaction with the α- or β-anomer, respectively, also seems surprising. However, stringent discrimination of anomeric carbohydrate antigens is well established in the protein-carbohydrate interactions of lectins (Sharon 1993). In addition to its immunological characterization, the linkage unit β-GlcAsn has been isolated from laminin of basement membranes and characterized by cochromatography with the authentic standard substance. Furthermore, the isolated compound yielded a FAB mass spectrum with a molecular ion mass peak as expected for a monohexosylasparagine, and was identical with the spectrum of the synthesized standard substance. In addition, by GLC, identical retention times were observed for the standard compound and the sample isolated from laminin.

These results together are taken as proof that the linkage unit β-GlcAsn exists in laminin from kidney. Laminin is an unusually large and complex glycoprotein (Sasaki et al. 1988). In addition to this novel N-glycosyl linkage it contains complex carbohydrates linked via the conventional N-linkage β-GlcNAcAsn. At this time we cannot quantitate the number of linkages of the β-GlcAsn type in the basement membrane protein, but our lowest estimate indicates a stoichiometry of 1 β-GlcAsn to 1 laminin. According to electrophoretic mobility, it seems to be the B2-chain of laminin that contains the novel linkage type.

We have obtained no immunological or chemical results that would suggest the presence of a linkage unit α-GlcAsn in protein extracts from mammalian kidney glomeruli.

We do not know at this time the mechanism of glycosylation of the extracellular glycoprotein laminin. However, it is possible that formation of the β-GlcAsn linkage, like conventional N-glycosylation, might occur in the lumen of the ER. In the cell surface glycoprotein from halobacteria, as in eukaryotic glycoproteins, all N-glucosyl sites show the consensus tripeptide structure Asn X Thr/Ser, and we would therefore expect this sequence as the site of β-GlcAsn-formation in laminin as well. It will be of interest to know the precise localization of this novel posttranslational modification activity, and to learn about the structural requirements that allow the N-glycosyltransferases for the conventional and the novel linkage unit to recognize their specific sites of operation.

In addition to its presence in isolated laminin, we have localized the epitope β-GlcAsn to the basement membranes of kidney tissue by immunofluorescence microscopy. Again, no indication was observed for the presence of a glucose residue in α-linkage to asparagine. For immunofluorescence microscopy we have worked out a method to chemically O-deglycosylate fixed tissue blocks. Surprisingly, under optimal deglycosylation conditions, the general architectural characteristics, as well as the immunogenic properties of the tissue, remain essentially intact. This method may be useful for the detection of masked epitopes in a more general way. Specifically, we would expect that peptide epitopes that are covered with carbohydrate may be efficiently unmasked. Likewise, this method may prove useful to discriminate whether a carbohydrate or a protein epitope is recognized by an antibody.

At present we do not know a possible specific function of the glycoconjugate linked to laminin by the novel unit β-GlcAsn. It seems remarkable, however, that the linkage is found on the cell surface of halobacteria as well as on the surface of mammalian cells, and in both cases in combination with glycosaminoglycans (the repeating unit sulfated saccharide in halobacteria (Paul and Wieland 1987), and the heparansulfate in the basement membrane). In addition, in both cell surfaces, identical disaccharides of the structure Glc (α 1–2) Gal are present in O-glycosyl linkage: linked to threonine in the halobacterial cell surface glycoprotein (Lechner and Wieland 1990) and to hydroxyproline in collagen type IV of the basement membrane (Levine and Spiro 1979). The surprising structural identities in two organisms as distant as halobacteria and mammals indicate that the novel structure described here might have an important function to establish or maintain the integrity of cell surface structures.

Acknowledgments. We are indebted to Dr. Timpl Martinsried, who provided us with generous amounts of laminin from EHS mouse tumor cells. We thank Dr. Schleicher, Munich, for the nonenzymatically glucosylated BSA and for the isolated glomeruli from rat and pig kidney, Dr. Lehmann, Heidelberg, for the FAB-MS in Fig. 6, and Dieter Jeckel, Klaus van Leyen, Mathias Neufeld, and Cordula Harter for critically reading the manuscript. Part of the figures are reproduced from the Journal of Cell Biology 1994, Vol. 124, No. 6 by copy right permisson of the Rockefeller University Press. This work was supported by the Deutsche Forschungsgemeinschaft (Wi 654/3-2).

References

Lechner J, Wieland F (1990) Structure and biosynthesis of procaryotic glycoproteins. Ann Rev Biochem 58:173–194

Levine MJ, Spiro RG (1979) Isolation from glomerular basement membrane of a glycopeptide containing both asparagine-linked and hydroxylysine-linked carbohydrate units. J Biol Chem 254:8121–8124

Paul G, Wieland F (1987) Sequence of the Halobacterial Glycosaminoglycan. J Biol Chem 262:9857–9593

Samokhin GP, Filimonov IN (1985) Coupling of peptides to protein carriers by mixed anhydride procedure. Anal Biochem 145:311–314

Sasaki M, Kleinman HK, Huber H, Deutzmann R, Yamada Y (1988) Laminin, a multidomain protein. J Biol Chem 263:16536–16544

Schreiner R, Schnabel E, Wieland F (1994) Novel N-Glycosylation in mammals: Laminin contains the linkage unit β-glycosylasparagine. J Cell Biol 124, 6

Sharon N (1993) Lectin-carbohydrate complexes of plants and animals: an atomic view. TIBS 18:221–226

Shibata S, Takeda T, Natori Y (1988) The structure of a Nephritogenoside (a nephritogenic glycopeptide with α-N-glycosidic linkage). J Biol Chem 263:12483–12485

Waldmann H, Kunz H (1983) Allylester als selektiv abspaltbare Carboxyschutzgruppe in der Peptid- und Glycopeptidsynthese. Liebigs Ann Chem 1983:1712–1725

Wieland F, Heitzer R, Schaefer W (1983) Asparaginylglucose: novel type of carbohydrate linkage. Proc Natl Acad Sci USA 80:5470–5474

Wieland F, Paul G, Sumper M (1985) Halobacterial flagellins are sulfated glycoproteins. J Biol Chem 260:15180–15185

Reprocessing of Membrane Glycoproteins

R. Tauber[1], B. Volz[1], W. Kreisel[2], N. Loch[3], G. Orberger[1], H. Xu[3], R. Nuck[3], and W. Reutter[3]

1 Introduction

Newly synthesized surface glycoproteins undergo extensive processing of their N-linked oligosaccharides during transport from the endoplasmic reticulum to the cell surface (for review Kornfeld and Kornfeld 1985). The fundamental processing reactions include removal of glucose and mannose residues from $Glc_3Man_9GlcNAc_2$ by glucosidases I and II and several mannosidases in the endoplasmic reticulum and the *cis*-Golgi to generate $Man_5GlcNAc_2$ which is the preferred substrate for N-acetylglucosaminyltransferase I of the medial Golgi compartment generating GlcNAc-$Man_5GlcNAc_2$. This structure can be acted upon by different enzymes including N-acetylglucosaminyltransferase III, N-acetylglucosaminyltransferase IV or can move from the medial to the *trans*-Golgi to form different types of hybrid structures. Alternatively, two additional mannose residues can be removed from GlcNAc-$Man_5GlcNAc_2$ by α-mannosidase II in the medial Golgi to form GlcNAc-$Man_3GlcNAc_2$, which can be acted upon by N-acetylglucosaminyltransferase II, III or IV, or by α6-fucosyltransferase, or can enter the *trans*-Golgi where the outer chains are extended by galactosyl-, fucosyl-, and sialyltransferases (Schachter 1986). Since the multiple processing enzymes may act upon the oligosaccharide substrates in various combinations, this nontemplate assembly line is perfectly suited to generate the wide variety of oligosaccharide structures found in secretory and membrane glycoproteins.

Whereas the biosynthetic pathway of glycoprotein glycans has been unraveled to some extent, little is known about the fate of the oligosaccharide chains once the glycoproteins have been transported through the glycosylating compartments and inserted into the surface membrane. This is particularly intriguing, since the majority of surface glycoproteins including receptors, ectoenzymes and cell adhesion molecules are rapidly internalized and recycled through endosomes, lysosomes, elements of the Golgi complex, or the *trans*-Golgi network (Goldstein et al. 1985; Woods et al. 1986; van Deurs et al. 1988; Stoorvogel et al. 1988; Griffiths et al. 1989; Gruenberg and Howell 1989; Brändli and Simons 1989; Green and Kelly 1992), where they may encounter glycosidases or glycosyltransferases either involved in oligosaccharide bio-

[1] Institut für Klinische Chemie und Biochemie, Universitätsklinikum Rudolf-Virchow, Freie Universität Berlin, Spandauer Damm 130, D-14050 Berlin, FRG.
[2] Medizinische Universitätsklinik der Albert-Ludwig-Universität Freiburg, D-79106 Freiburg, FRG.
[3] Institut für Molekularbiologie und Biochemie, Freie Universität Berlin, Arnimallee 22, D-14195 Berlin, FRG.

44. Colloquium Mosbach 1993
Glyco- and Cellbiology

synthesis or degradation, or of still unknown function (Snider and Rogers 1985, 1986; Kreisel et al. 1988, 1993; Reichner et al. 1988; Duncan and Kornfeld 1988; Brändli and Simons 1989). Since the oligosaccharide units of surface glycoproteins are potential substrates for these enzymes, the question arises as to whether the oligosaccharide chains of surface glycoproteins may undergo further processing by these enzymes during intracellular trafficking. We have addressed this question in cultured rat hepatocytes, rat and human hepatoma cells, and Chinese hamster ovary (CHO) cells using three experimental approaches:

1. As a measure of a possible trimming of glycoprotein glycans by removal of single sugar residues from the oligosaccharides, half-lives of the different sugar residues were determined and compared to the half-live of the polypeptide backbone of the glycoproteins.
2. In order to identify trimming intermediates, structures of N-linked oligosaccharides of surface glycoproteins were analyzed at different times after insertion of the glycoproteins into the surface membrane.
3. Reglycosylation of surface glycoproteins during return from the cell surface to glycosylating compartments was examined comparing trafficking to the *cis*-Golgi, the medial/*trans*-Golgi and the *trans*-Golgi network.

Surface glycoproteins of different membrane orientation, glycosylation and function were studied including the transferrin receptor (Orberger et al. 1992), the serine peptidase dipeptidylpeptidase IV (DPP IV) (Loch et al. 1992), 6D-cadherin, a cell adhesion molecule of the cadherin family (Berndorff et al. submitted), and gp110/cell-CAM 105, a member of the Ig-supergen family having functions both as a cell adhesion molecule and as a bile salt transporter (Müller et al. 1991; Becker et al. 1993).

2 Reprocessing of Surface Glycoproteins by Trimming of the N-Linked Oligosaccharides

2.1 Half-Lives of Sugar Residues in Surface Glycoproteins

The majority of surface glycoproteins of mammalian cells have comparably long half-lives ranging from approximately 20 to 90 h (for review Hare 1990). During that period, surface glycoproteins are either resident in the plasma membrane or cycling in between the plasma membrane and intracellular compartments. In order to trace if the oligosaccharide chains of the glycoproteins are subject to trimming during that period of the glycoprotein life span, half-lives of the different sugar residues in the glycoproteins were measured in rat liver in vivo in pulse-chase experiments after metabolic radiolabeling with sugar precursors (Kreisel et al. 1980; Tauber et al. 1983; Volk et al. 1983). The different sugar residues exhibited a distinct graduation of their half-lives that was related to the position of each sugar within the structure of complex N-linked oligosaccharides (Table 1). Sugar residues in terminal or penultimate positions, L-fucose, N-acetylneuraminic acid and D-galactose had the shortest half-lives, only 1/6 to 1/3 that of the protein backbone. By contrast, the core sugar D-mannose had

Table 1. Half-lives of sugar residues and protein moieties of plasma membrane glycoproteins. Half-lives were determined in rat liver (gp60, gp80, gp120, gp140, gp160, DPP IV), primary cultures of rat hepatocytes (ASGP-R) or HepG2 cells (TfR). Polypeptide moieties were labeled with either $^{+}$L-[4,5-^{3}H] leucine or *L-[^{35}S] methionine

Half-life (h) of						
Glycoprotein	L-[^{3}H]Fuc	[^{3}H]NeuAc	D-[^{3}H]Gal	D-[^{3}H]GlcNAc	D-[^{3}H]Man	Polypeptide
gp60[a]	21	33	n.d.	n.d.	58	62^{+}
gp80[a]	17	31	n.d.	n.d.	38	85^{+}
DPP IV[b,c]	12,5	33	20	43	58	70^{*}
gp120[a]	21	27	n.d.	n.d.	51	88^{+}
gp140[a]	16	29	n.d.	n.d.	66	78^{+}
gp160[a]	16	26	n.d.	n.d.	26	52^{+}
ASGP-R[d]	12	n.d.	n.d.	n.d.	n.d.	24^{*}
TfR[e]	14	n.d.	11	n.d.	12	14^{*}

Data from [a] Tauber et al. 1983; [b] Kreisel et al. 1983; [c] Volk et al. 1983; [d] Abetz et al. in prep. [e] Orberger et al. in prep.

half-lives similar to that of the protein backbone in most of the glycoproteins investigated (gp60, DPP IV, gp140). In some other glycoproteins, D-mannose had shorter half-lives, either closely related to that of the terminal sugars (gp160) or in between that of the protein backbone and that of the terminal sugars (gp80, gp120). These data indicated that during the life span of cell surface glycoproteins, the sugar residues may be stepwise removed from the nonreducing end of the oligosaccharides partly including even the core region of the glycans.

Rapid loss of sugar residues could also be shown in plasma membrane DPP IV of cultured rat hepatocytes using a different experimental approach (Kreisel et al. 1988). DPP IV metabolically labeled with L-[^{3}H]fucose, D-[^{3}H]mannose, or L-[^{35}S]methionine was tagged with monospecific antibodies when exposed on the cell surface. At different times thereafter, the immune complex was isolated by binding to protein A-Sepharose. During the experiment the immune complex was stable, as proven by the constant specific radioactivity of the DPP IV-antibody complex after labeling of DPP IV with L-[^{35}S]methionine (Fig. 1). In accordance with the results obtained in intact liver in vivo, L-[^{3}H]fucose was rapidly lost from DPP IV, whereas D-[^{3}H]mannose remained bound to the glycoprotein. Preferential loss of terminal sugars has also been reported for surface glycoproteins of rat HTC and Reuber H-35 cells, and for membrane glycoproteins transferred into the plasma membrane of mouse fibroplasts using fusion with reconstituted phospholipid vesicles (Baumann et al. 1983).

In order to estimate a possible kinetic relationship of the loss of terminal sugars with the rate of endocytosis and recycling, the metabolic stability of sugar residues

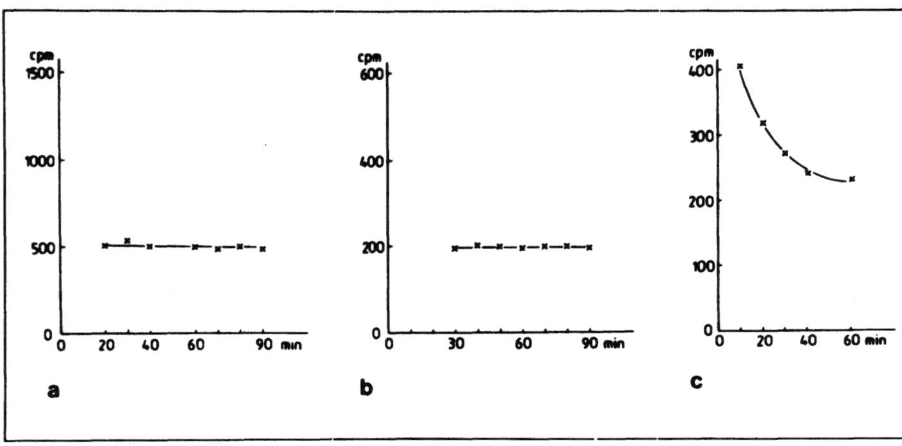

Fig. 1a–c. Decay of specific radioactivity of plasma membrane DPP IV in primary cultures of rat hepatocytes. Cells were pulse-chase labeled with L-[^{35}S]methione (**a**), D-[2–^{3}H]mannose (**b**), or L-[6–^{3}H]fucose (**c**) at 37 °C, and were than incubated with a monovalent anti-DPP IV antibody for 2 h at 4 °C. After removal of unbound antibody by a washing procedure, cells were rewarmed to 37 °C and were further incubated for the times indicated in the presence of excess of unlabeled precursor. At the end of the incubation period, cells were extracted with Triton X-100/Tris buffer. The immune complex was bound to protein A-Sepharose and counted for radioactivity. (Data from Kreisel et al. 1988)

was measured in the asialoglycoprotein receptor of primary cultured rat hepatocytes and in the transferrin receptor of Hep G2 cells (Table 1). In neither one of the two endocytic receptors which are rapidly internalized in both cell types with a half-life of internalization of 5 to 10 min, sugar residues were preferentially removed from the oligosaccharide moiety, but were turning over with approximately the same half-lives as the receptor polypeptides. These results show that removal of terminal sugars is not directly related to endocytosis and to the control of receptor recycling. Moreover, according to these results, distinct surface glycoproteins may escape reprocessing, indicating that reprocessing is a selective process.

2.2 Structural Identification of Trimmed Oligosaccharides on Surface Glycoproteins

In order to trace the generation of trimmed oligosaccharides on cell surface glycoproteins, oligosaccharide structures of dipeptidylpeptidase IV, 6D-cadherin, gp110/cell-CAM 105, and transferrin receptor were analyzed at different times after insertion of these glycoproteins into the surface membrane. Several studies (Snider and Rogers 1985; Kreisel et al. 1988, 1993; Reichner et al. 1988; Duncan and Kornfeld 1988; Brändli and Simons 1989) had shown that oligosaccharides of the N-linked complex type of surface glycoproteins may be resialylated and refucosylated during

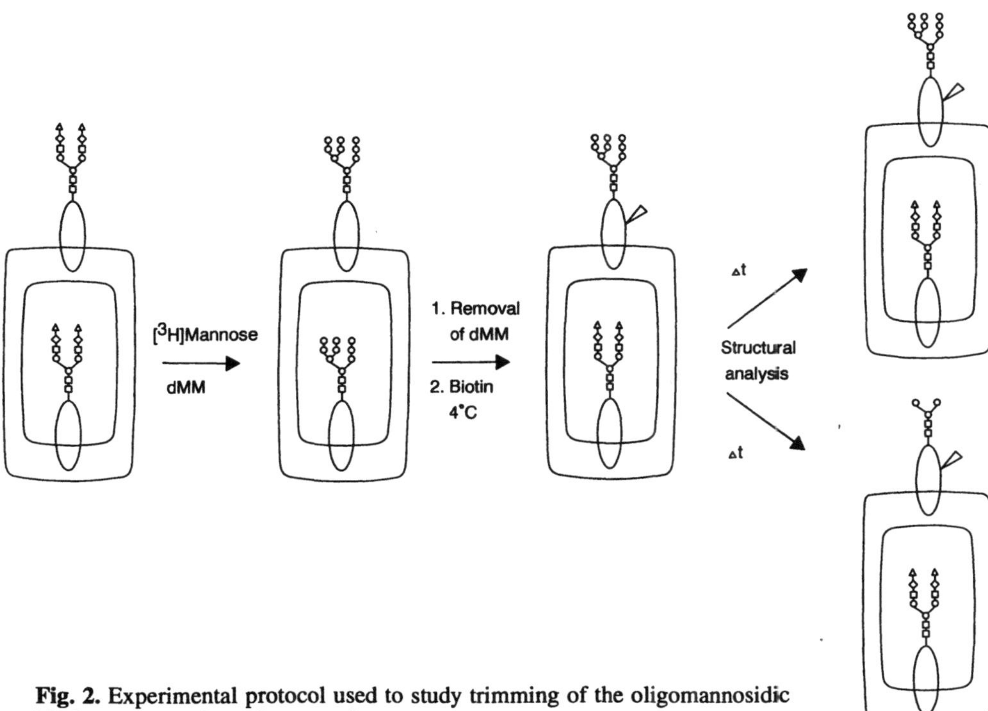

Fig. 2. Experimental protocol used to study trimming of the oligomannosidic N-glycans on surface glycoproteins

internalization and recycling. This repair mechanism would conceal trimming of complex type structures. In order to circumvent this problem, oligosaccharides of the high mannose type were monitored for loss of terminal mannose residues. Repair of this type of N-linked oligosaccharides is highly unlikely, since no mechanism has so far been described in mammalian cells to mediate the transfer of mannose residues to mature glycoproteins. As shown in Fig. 2, the formation of surface glycoproteins glycosylated with oligomannosidic structures was induced by inhibition of processing mannosidase I with deoxymannojirimycin. After transport to the surface, the glycoproteins were covalently labeled with biotin residues allowing these glycoproteins to be isolated by affinity chromatography on streptavidin agarose at different times thereafter (Busch et al. 1989). Analysis of the N-linked oligosaccharides showed that immediately after transport to the surface membrane, the glycoproteins were almost completely glycosylated with $Man_9GlcNAc_2$ structures in accordance with the reported inhibitory effect of deoxymannojirimycin. During the life span of the glycoproteins, $Man_9GlcNAc_2$ structures of dipeptidylpeptidase IV and gp110/cell-CAM 105 were trimmed to smaller structures with sizes of $Man_8GlcNAc_2$, $Man_7GlcNAc_2$, $Man_6GlcNAc_2$, and $Man_5GlcNAc_2$ in relative amounts of $Man_5GlcNAc_2$ > $Man_6GlcNAc_2$ > $Man_7GlcNAc_2$ > $Man_8GlcNAc_2$ (Loch et al. in prep.). No structures smaller than $Man_5GlcNAc_2$ could be detected. By contrast to dipeptidylpeptidase IV and gp110/cell-CAM 105, both transferrin receptor and 6D-cadherin retained $Man_9GlcNAc_2$ structures throughout their lifetime in accordance with the results of the turnover studies (Table 1). Trimming of the oligomannosidic N-glycans of dipeptidylpeptidase IV and gp110/cell-CAM 105 was blocked when cells were recultured in the presence of deoxymannojirimycin. These results show that N-glycans of

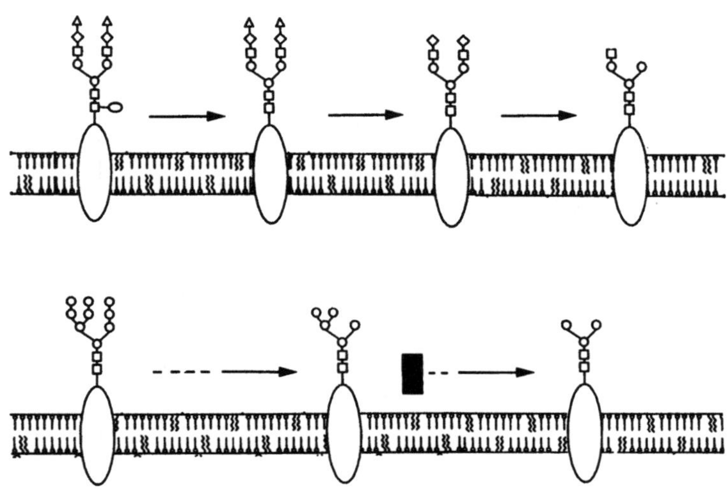

Fig. 3. Trimming of oligosaccharides of the complex and of the oligomannosidic type of cell surface glycoproteins

selected surface glycoproteins are subject to a stepwise demannosylation by mannosidases sensitive to deoxymannojirimycin (Fig. 3).

2.3 Subcellular Localization of Trimming Reactions

The cellular compartments and the glycosidases involved in the trimming of surface glycoproteins have not yet been identified. Preliminary experiments have shown that chloroquine does not affect the loss of L-fucose from plasma membrane glycoproteins while significantly inhibiting degradation of the polypeptide portion of these proteins (Tauber et al. 1985). This indicated that defucosylation occurs extralysosomally. This assumption is further supported by the observation that demannosylation is not inhibited by swainsonine at concentrations known to inhibit lysosomal mannosidases (Loch et al. in prep.).

According to the present kinetic and structural data, selected surface glycoproteins undergo trimming of their oligosaccharides by removal of sugar residues from the nonreducing end of both complex and high mannose-type glycans (Fig. 3). Trimming of complex-type glycans particularly includes L-fucose, sialic acid and D-galactose residues, whereas in most cases N-actylglucosamine and mannose residues are not affected. N-Glycans of the oligomannosidic type are trimmed to $Man_5GlcNAc_2$ oligosaccharides, but not to smaller structures. The subcellular localization of the trimming reactions is unknown. Preliminary experiments indicate that trimming occurs outside lysosomes.

3 Reglycosylation of Surface Glycoproteins During Return to Compartments of the Biosynthetic Route

Internalized plasma membrane glycoproteins have been shown to return to elements of the biosynthetic route including *cis*-Golgi, medial/*trans*-Golgi (Woods et al. 1986) and the *trans*-Golgi network (van Deurs et al. 1988). Since these compartments contain various glycosyltransferases, it has been proposed that sugar residues may be transferred to surface glycoproteins even during internalization and recycling. This has been clearly shown for the transfer of sialic acid residues by sialyltransferases of the *trans*-Golgi network employing surface glycoproteins, which were experimentally rendered substrates for sialyltransferases either by exogalactosylation of surface glycoproteins in mutant cell lines with impaired galactosylation (Duncan and Kornfeld 1988; Brändli and Simons 1989; Green and Kelly 1990), or by surface desialylation (Snider and Rogers 1985; Reichner et al. 1988). It remained unknown, however, whether this post-biosynthetic modification of the oligosaccharides of internalized surface glycoproteins is limited to transfer of sialic acid residues or may also include transfer of other sugars or even the reprocessing of nonprocessed high mannose-type oligosaccharides to complex structures.

Reglycosylation of cell surface glycoproteins during internalization and recycling was studied by example of DPP IV in cultured rat hepatocytes (Kreisel et al. 1988,

1993), and by example of DPP IV and the transferrin receptor in human hepatocarcinoma Hep G2 cells (Volz et al. in prep.). In order to discriminate between newly synthesized glycoproteins passing through the glycosylating compartments during biosynthesis, and glycoproteins that return to these compartments from the cell surface, plasma membrane proteins were surface labeled either covalently or noncovalently. In the study employing rat hepatocytes, DPP IV molecules exposed on the cell surface were tagged with a monospecific antibody at 4 °C. After removal of excess antibody, cells were rewarmed to 37 °C to allow DPP IV/antibody complexes to be internalized and to recycle. In Hep G2 cells surface proteins were labeled with the cleavable biotinylation reagent NHS-SS-biotin. Internalized surface glycoproteins were then recovered from intracellular compartments either by binding to protein A-Sepharose or by affinity chromatography on strepavidin-agarose. Return to sialyltransferases localized predominantly in the *trans*-Golgi network (Roth et al. 1985) was studied in hepatocytes measuring the incorporation of N-[^{3}H]acetylneuraminic acid using N-[^{3}H]acetylmannosamine as a precursor (Kreisel et al. 1988, 1993). In Hep G2 cells, surface proteins were desialylated with neuraminidase at 4 °C beforehand in order to increase acceptor sites for resialylation. Resialylation was analyzed by isoelectric focusing of the immunoadsorbed surface glycoproteins (Volz et al. in prep.). Whereas transfer of sialic acid residues to surface DPP IV could be demonstrated by both of the two experimental approaches, no resialylation of the transferrin receptor and, likewise, of receptor-bound asialotransferrin was detectable. Recycling to fucosyltransferases localized predominantly in medial/*trans*-elements of the Golgi complex, was traced measuring incorporation of L-[^{3}H]fucose into surface-labeled proteins. Several cell surface glycoproteins, including DPP IV but not the transferrin receptor, acquired L-fucose during recycling. The present data show that terminal sugars may be transferred to a restricted set of surface glycoproteins even subsequent to biosynthesis. These results are in line with previous reports that a restricted set of surface proteins may be resialylated while recycling through the *trans*-Golgi network (Duncan and Kornfeld 1988; Reichner et al. 1988; Brändli and Simons 1989). On the other hand, reports on glycoprotein recycling to galactosyltransferases of the Golgi complex are controversial. Return to sialyltransferases, but not to galactosyltransferases has been observed for the 215-kDa and the 46-kDa mannose-6-phosphate receptors (Duncan and Kornfeld 1988). By contrast, Huang and Snider (1993) have reported that the cation-independent mannose-6-phosphate/insulin-like growth factor-II receptor recycles to both galactosyl- and sialyltransferase in the ldlD mutant of Chinese hamster ovary cells.

Whereas return to the *trans*-Golgi network has been shown by several groups using different experimental approaches, return to inner elements of the biosynthetic route is a matter of controversy. In these compartments recycling glycoproteins might encounter enzymes involved in early steps of oligosaccharide processing. Return to mannosidase I localized in *cis*-Golgi elements has been reported for the transferrin receptor in erythroleukemic K562 cells (Snider and Rogers 1986). However, this finding could not be confirmed by Neffjes et al. (1988) studying the transferrin receptor and MHC class I antigens in the same cell line. Likewise, no significant return of the 215-kDa and the 46-kDa mannose-6-phosphate receptors to mannosidase I in *cis*-Golgi cisternae could be detected (Duncan and Kornfeld 1988).

With respect to these conflicting results, recycling of transferrin receptor and DPP IV to *cis*-Golgi elements was studied in Hep G2 cells using surface biotinylation to tag both proteins on the cell surface. In accordance with the results of Neefjes et al. (1988), we could not demonstrate recycling of either one of both glycoproteins to processing enzymes of the *cis*-Golgi (Volz et al. in prep.), indicating that surface glycoproteins may return to medial/*trans*-elements of the Golgi complex, but not to *cis*-Golgi cisternae.

4 Factors Influencing the Reprocessing of Surface Glycoproteins

The influence of cell proliferation and transformation on the rates and the extent of oligosaccharide reprocessing was studied in regenerating liver and in hepatoma. In proliferating liver after partial hepatectomy, the half-life of L-fucose was extended up to threefold in DPP IV and in other plasma membrane glycoproteins, whereas half-lives of D-mannose and N-acetylneuraminic acid were not significantly altered (Kreisel et al. 1984; Tauber et al. 1989). This indicates that in proliferating liver selectively the removal of L-fucose from cell surface glycoproteins is reduced. By contrast, in hepatoma both L-fucose and D-mannose had similar short half-lives in plasma membrane glycoproteins, indicating that not only peripheral but also core sugars of the oligosaccharides are preferentially split off from surface glycoproteins (Tauber et al. 1989). Mechanisms responsible for alterations of oligosaccharide reprocessing might include altered expression of glycosidases or glycosyltransferases, different rates and routes of intracellular transport, and the formation of oligosaccharide structures with different sensitivity to reprocessing enzymes.

5 Perspectives

In conjunction with the data demonstrating trimming of the oligosaccharides of surface glycoproteins, the results on reglycosylation support the concept that plasma membrane glycoproteins may undergo continuous reprocessing of their oligosaccharides by partial de- and reglycosylation, as schematically shown in Fig. 4. Many questions about oligosaccharide reprocessing of plasma membrane glycoproteins remain to be answered. Which glycosidases and glycosyltransferases are involved? Does reprocessing affect each of the different glycosylation sites of a glycoprotein, or is partial de- and reglycosylation restricted to distinct oligosaccharide units? By analogy to oligosaccharide processing during biosynthesis, it is conceivable that the different oligosaccharide units of a glycoprotein may be reprocessed to a different extent, determined by their steric accessibility to glycosidases or glycosyltransferases. Is reprocessing specific for plasma membrane glycoproteins, or do glycoproteins of other cellular compartments undergo similar modifications of their oligosaccharide moiety?

The biological significance of oligosaccharide reprocessing remains to be established. Trimming and reglycosylation of surface glycoproteins could reflect the occa-

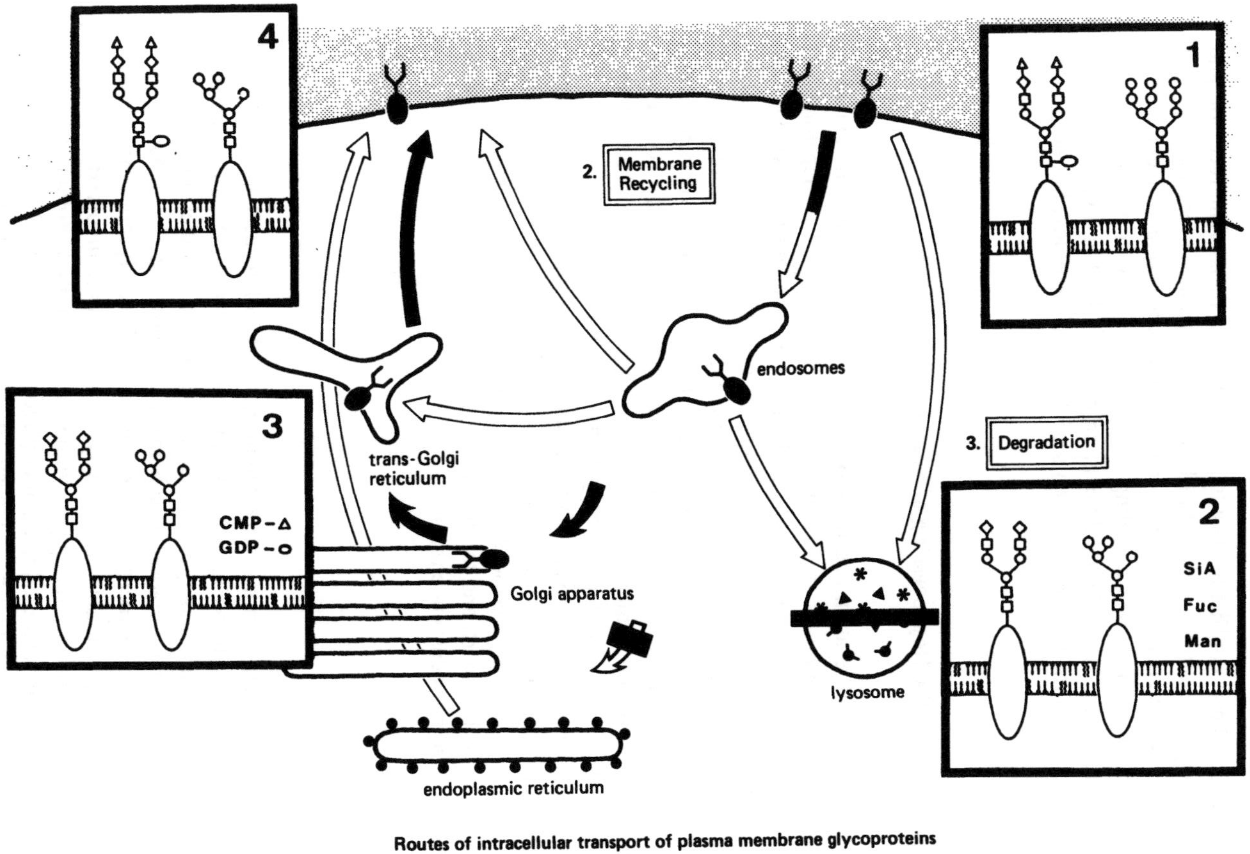

Fig. 4. Oligosaccharide reprocessing of cell surface glycoproteins. *1*, *2* Surface glycoproteins undergo trimming by removal of outer sugar residues, e.g., of fucose and sialic acid residues from oligosaccharides of the complex type, or of mannose residues from oligomannosidic structures. *3*, *4* Partial deglycosylation is followed by transfer of sialic acid and fucose residues to oligosaccharides of the complex type of glycoproteins routed through medial/*trans* elements of the Golgi complex and the *trans*-Golgi network before return to the cell surface

sional loss of outer sugar residues and subsequent repair. On the other hand, reprocessing might be a regulated process in cellular adaptation, conferring structural oligosaccharide modifications to a restricted set of surface glycoproteins in response to cellular or extracellular stimuli.

According to the present data, trimming of surface glycoproteins occurs either on the cell surface or during endocytosis, while reglycosylation takes place during recycling through compartments of the biosynthetic pathway. Whereas both the routes of biosynthesis and endocytosis have been studied in detail, little is known about the regulation and the biological function of the routes connecting these two pathways. The observation that reglycosylation is selective for distinct surface glycoproteins indicates that recycling from the endocytic route to glycosylating compartments is directed by signal structures present on these proteins.

Acknowledgments. This work was supported by the Deutsche Forschungsgemeinschaft (Sonderforschungsbereiche 154 and 312, and Re 523/3-3) and by the Maria-Sonnenfeld-Gedächtnisstiftung.

References

Baumann H, Hou E, Jahreis GP (1983) Preferential degradation of the terminal carbohydrate moiety of membrane glycoproteins in rat hepatoma cells and after transfer to the membranes of mouse fibroblasts. J Cell Biol 96:139–150

Becker A, Lucka L, Kilian C, Kannicht C, Reutter W (1993) Characterization of the ATP-dependent taurocholate-carrier protein (gp110) of the hepatocyte canalicular membrane. Eur J Biochem 214:539–548

Brändli AW, Simons K (1989) A restricted set of apical proteins recycle through the *trans*-Golgi network in MDCK cells. EMBO J 8:3207–3213

Busch G, Hoder D, Reutter W, Tauber R (1989) Selective isolation of individual cell surface proteins from tissue culture cells by a cleavable biotin label. Eur J Cell Biol 50:257–262

van Deurs B, Sandvig K, Peterson OW, Olsnes S, Simons K, Griffiths G (1988) Estimation of the amount of internalized ricin that reaches the *trans*-Golgi network. J Cell Biol 106:253–267

Duncan JR, Kornfeld S (1988) Intracellular movement of two mannose-6-phosphate receptors: return to the Golgi apparatus. J Cell Biol 106:617–628

Goldstein JL, Brown MS, Anderson RGW, Russell DW, Schneider WJ (1985) Receptor-mediated endocytosis: concepts emerging from the LDL receptor system. Annu Rev Cell Biol 1:1–39

Green SA, Kelly RB (1990) Endocytotic membrane traffic to the Golgi apparatus in a regulated secretory cell line. J Biol Chem 265:21269–21278

Green SA, Kelly RB (1992) Low density lipoprotein receptor and cation-independent mannose-6-phosphate receptor are transported from the cell surface to the Golgi apparatus at equal rates in PC12 cells. J Cell Biol 117:47–55

Griffiths G, Back R, Marsh M (1989) A quantitative analysis of the endocytic pathway in baby hamster kidney cells. J Cell Biol 109:2703–2720

Gruenberg J, Howell KE (1989) Membrane traffic in endocytosis: insights from cell free assays. Annu Rev Cell Biol 5:453–481

Hare JF (1990) Mechanisms of membrane protein turnover. Biochim Biophys Acta 1031:71–90

Huang KM, Snider MD (1993) Glycoprotein recycling to the galactosyltransferase compartment of the Golgi complex. J Biol Chem 268:9302–9310

Kornfeld R, Kornfeld S (1985) Assembly of asparagine-linked oligosaccharides. Annu Rev Biochem 54:631–664

Kreisel W, Volk BA, Büchsel R, Reutter W (1980) Different half-lives of the carbohydrate and protein moieties of a 110 000-dalton glycoprotein isolated from plasma membranes of rat liver. Proc Natl Acad Sci USA 77:1828–1831

Kreisel W, Reutter W, Gerok W (1984) Modification of the intramolecular turnover of terminal carbohydrates of dipeptidylaminopeptidase IV isolated from rat liver plasma membrane during liver regeneration. Eur J Biochem 138:435–438

Kreisel W, Hanski C, Tran-Thi T-A, Katz N, Decker K, Reutter W, Gerok W (1988) Remodeling of a rat hepatocyte plasma membrane glycoprotein. J Biol Chem 263:11736–11742

Kreisel W, Hildebrand H, Mössner W, Tauber R, Reutter W (1993) Oligosaccharide reprocessing and recycling of a cell surface glycoprotein in cultured rat hepatocytes. Biol Chem Hoppe-Seyler 374:255–263

Loch N, Tauber R, Becker A, Hartel-Schenk S, Reutter W (1992) Biosynthesis and metabolism of dipeptidylpeptidase IV in primary cultured rat hepatocytes and Morris hepatoma 7777 cells. Eur J Biochem 210:161–168

Müller M, Ishikawa T, Berger U, Klünemann C, Lucka L, Schreyer A, Kannicht C, Reutter W, Kurz G, Keppler D (1991) ATP-dependent transport of taurocholate across the hepatocte canalicular membrane mediated by a 110-kDa glycoprotein binding ATP and bile salt. J Biol Chem 266:18920–18926

Orberger G, Geyer R, Stirm S, Tauber R (1992) Structure of the N-linked oligosaccharides of the human transferrin receptor. Eur J Biochem 205:257–267

Reichner JS, Whiteheart SW, Hart GW (1988) Intracellular trafficking of cell surface sialoglycoconjugates. J Biol Chem 263:16316–16326

Roth J, Taatjes DJ, Lucocq JM, Weinstein J, Paulson JC (1985) Demonstration of an extensive *trans*-tubular network continuous with the Golgi apparatus stack that may function in glycosylation. Cell 43:287–295

Schachter H (1986) Biosynthetic controls that determine the branching and microheterogeneity of protein-bound oligosaccharides. Biochem Cell Biol 64:163–181

Snider MD, Rogers OC (1985) Intracellular movement of cell surface receptors after endocytosis: resialylation of asialotransferrin receptor in human erythroleukemia cells. J Cell Biol 100:826–834

Snider MD, Rogers OC (1986) Membrane traffic in animal cells: cellular glycoproteins return to the site of Golgi mannosidase I. J Cell Biol 103:265–275

Stoorvogel W, Geuze HJ, Griffith JM, Strous GJ (1988) The pathways of endocytosed transferrin and secretory protein are connected in the *trans*-Golgi reticulum. J Cell Biol 106:1821–1829

Tauber R, Park CS, Reutter W (1983) Intracellular heterogeneity of degradation of plasma membrane glycoproteins – evidence for a general characteristic. Proc Natl Acad Sci USA 80:4026–4029

Tauber R, Heinze K, Reutter W (1985) Effect of chloroquine on the degradation of L-fucose and the polypetide moiety of plasma membrane glycoproteins. Eur J Cell Biol 39:380–385

Tauber R, Kronenberger CH, Reutter W (1989) Decreased intramolecular turnover of L-fucose in membrane glycoproteins of rat liver during liver regeneration. Biol Chem Hoppe-Seyler 370:1221–1228

Tauber R, Park CS, Becker A, Geyer R, Reutter W (1989) Rapid intramolecular turnover of N-linked glycans in plasma membrane glycoproteins. Eur J Biochem 186:55–62

Volk B, Kreisel W, Köttgen E, Gerok W, Reutter W (1983) Heterogeneous turnover of terminal and core sugars within the carbohydrate chain of dipeptidylaminopeptidase IV isolated from rat liver plasma membrane. FEBS Lett 163:150–152

Woods JW, Doriaux M, Farquhar MG (1986) Transferrin receptors recycle to *cis* and middle as well as *trans*-Golgi cisternae in Ig-secreting myeloma cells. J Cell Biol 103:277–286

Mutational Analysis of Carbohydrate and Phospholipid Modifications of a Cell Adhesion Protein

G. Gerisch[1], J. Faix[1], E. Wallraff[1], A. A. Noegel[1], A. Barth[1], R. Lützelschwab[1], M. Westphal[1], G. Langanger[1], and D. Francis[2]

1 Introduction

The contact site A (csA) glycoprotein is a developmentally regulated cell adhesion molecule (Beug et al. 1973; Müller and Gerisch 1978; Noegel et al. 1985). It is absent from growth-phase cells, and is maximally expressed at the aggregation stage of *Dictyostelium discoideum*, i.e., during transition from the state of single cells to a multicellular organism (Murray et al. 1981). According to its sequence, the csA protein consists of three regions: a large extracellular N-terminal domain, a proline-, serine- and threonine-rich sequence which resembles the hinge region of immunoglobulins, and a carboxyterminal stretch of hydrophobic amino acids (Noegel et al. 1986). The protein is cotranslationally modified by N-linked carbohydrate residues at the N-terminal domain, and by a ceramide-based phospholipid (PL) anchor (Stadler et al. 1989). The pro/ser/thr-rich stretch of the C-terminal region near to the plasma membrane is decorated with O-linked carbohydrate residues during passage of the protein through the Golgi apparatus, and the N-linked carbohydrate residues are sulfated during the passage (Hohmann et al. 1985; Hohmann et al. 1987b).

In this chapter we discuss four aspects of the cell-adhesion system of *D. discoideum* and of a related species, *Polysphondylium pallidum*: (1) changes in adhesion and morphogenesis after overexpression or elimination of the csA glycoprotein; (2) presence of another PL-anchored protein that may account for redundancy in the cell-adhesion system; (3) consequences of eliminating O- and N-linked carbohydrate residues or of replacing the PL-anchor by a transmembrane domain; (4) species specificity of adhesion and a morphogenetic role of protein-linked carbohydrates in *Polysphondylium.*

2 Overexpression of the csA-Glycoprotein Enables Growth-Phase Cells to Aggregate and Leads to the Induction of Developmentally Regulated Proteins

To specify the function of csA in cell adhesion and to establish a role in morphogenesis, the protein has been selectively eliminated by mutation or has been overexpressed under the control of two different promoters. The transcription from its own csA promoter begins after induction of development at about 3 h of starvation, and is

[1] Max-Planck-Institut für Biochemie, D-82152 Martinsried, FRG.
[2] Permanent Adress: School of Life and Health Sciences, University of Delaware, Newark, DE 19716, USA.

44. Colloquium Mosbach 1993
Glyco- and Cellbiology

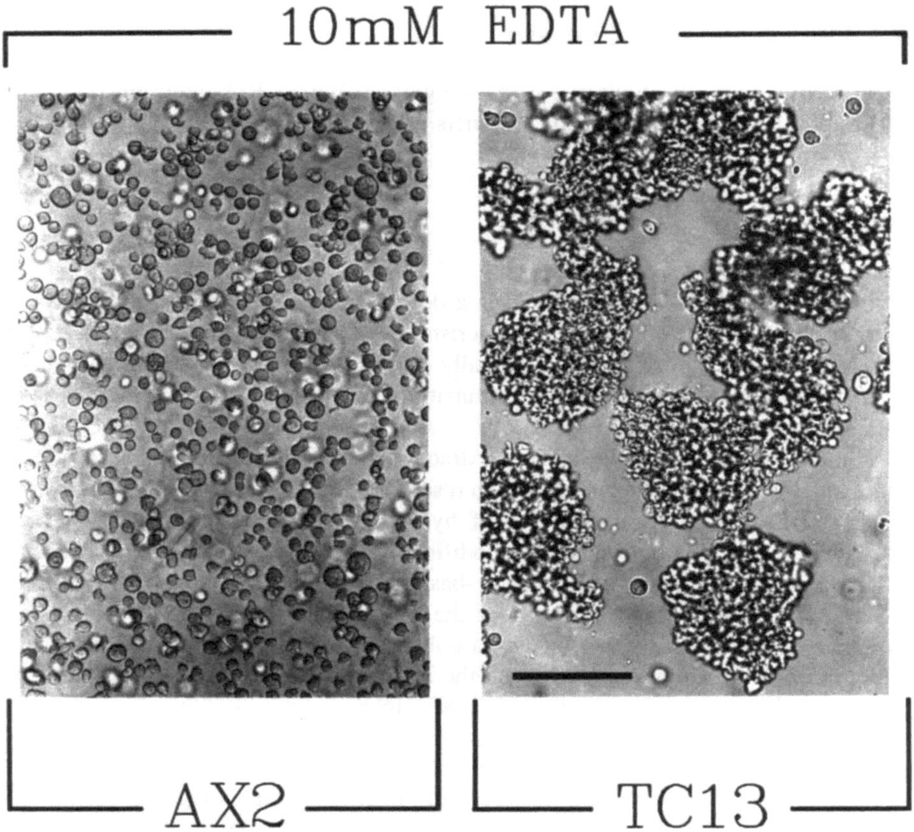

Fig. 1. EDTA-stable adhesion in cell suspension induced by overexpression of the csA protein. *Left* Control of growth-phase cells of wild-type AX2 in which csA is not expressed; *right* growth-phase cells of transformant TC13 which strongly expresses csA under the control of a constitutive promoter. *Bar* 100 μm

strongly enhanced by periodic stimulation of the cells with cAMP pulses (Faix et al. 1992; Desbarats et al. 1992). Therefore, this promoter can be used for the regulated expression of the protein. The actin15 promoter, on the other hand, is already active in growth-phase cells, and can thus be used to study the function of normal and mutated csA while the endogenous csA promoter is silent.

When expressed in transformed cells during the growth phase, the csA protein becomes fully decorated with N- and O-linked oligosaccharides and is anchored in the plasma membrane by phospholipid (Faix et al. 1990). Expression of this single protein in addition to the membrane components normally present during growth enables the cells to strongly agglutinate in the presence and absence of EDTA (Fig. 1). The formation of EDTA-stable cell contacts is a standard assay for csA activity (Bozzaro et al. 1987). More importantly, when growing on a plastic surface, cells of the csA-expressing transformants aggregate into blastula-like spheres (Fig. 2). They are also

TC13

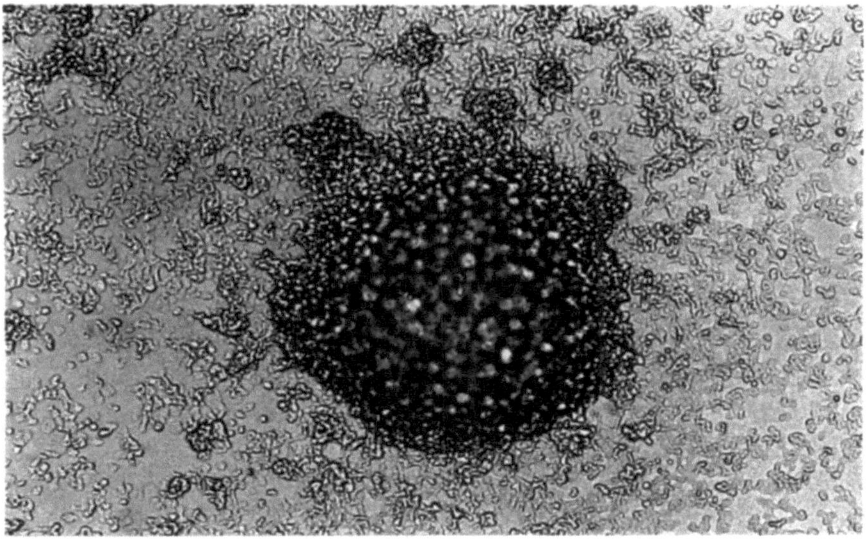

AX2

Fig. 2. CsA expression enables TC13 cells to aggregate into streams and to form hollow spheres on a polystyrene surface in the presence of nutrient medium *(top)*. Under the same conditions, untransformed cells of wild-type AX2 remain dispersed on the surface *(bottom)*. *Bar* 100 μm

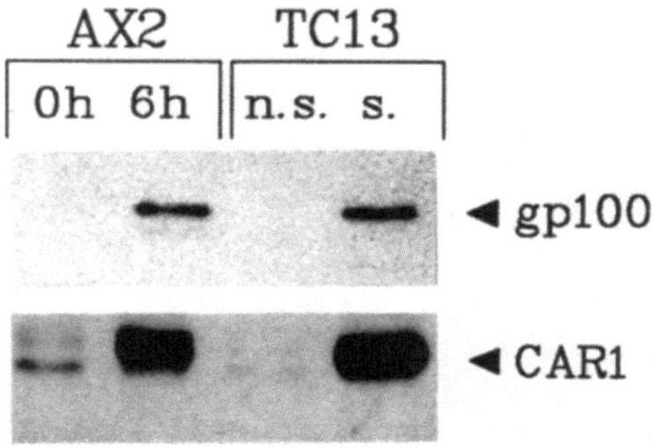

Fig 3. Expression of two developmentally regulated genes as a result of csA expression during growth of transformant TC13. In wild-type AX2, gp100 is a strictly developmentally regulated protein of unknown function that is not expressed during growth (0 h) (Müller-Taubenberger 1989), CAR1 is a cAMP receptor which like gp100 is maximally expressed in aggregating cells (6 h) (Saxe III et al. 1988). In TC13 both proteins are not expressed at an early stage of growth in cells that have not yet started to stream *(n.s.)*, while they are detected later during growth when they stream together *(s.)* as in Fig. 2. Antibodies used for the immunoblots: anti-CAR1, courtesy of Dr. Peter Devreotes, anti gp100 of Dr. Annette Müller-Taubenberger

induced to express CAR1 cAMP-receptors and other marker proteins of the aggregation stage (Fig. 3), which normally are only expressed in starving cells (Faix 1993).

3 CsA Elimination Demonstrates Redundancy in the Cell Adhesion System of Aggregating Cells

A mutant lacking the csA glycoprotein has been selected after chemical mutagenesis by use of a fluorescence activated cell sorter and a csA-protein specific antibody (Noegel et al. 1985). Cells of this mutant agglutinate very weakly in the presence of EDTA when tested in an agitated suspension after 6 h of starvation, i.e., at the aggregation-competent stage. Nevertheless, when these cells aggregate on a solid agar surface where they are not exposed to shear forces, they sufficiently adhere to each other to enter the multicellular stage and to complete development. This result shows the presence of a second adhesion system in aggregating cells, which acts independently of csA. In search for a candidate glycoprotein that enables csA-null cells to aggregate, we have purified a glycoprotein with an apparent molecular mass of 130 kDa (gp130) which resembles csA in three respects: (1) it incorporates inositol and ethanol, two components of a phospholipid anchor (Stadler et al. 1989); (2) its N-linked carbohydrate residues are strongly sulfated and intensely labeled by ^{3}H-L-fucose; and (3) gp130 reacts as the only other glycoprotein with a particular monoclonal antibody against O-linked carbohydrate residues of the csA molecule. According to the mAb 19-14-19 label, gp130 is already expressed at a low level in growth-phase cells. In starving cells, labeling with the antibody is increased, but this increase is suppressed by pulsatile cAMP signals rather than enhanced as in the case of csA. Based on

```
        10        20        30        40        50        60
gp130   VMSDLLFNLYGYDKSLDPCNSNSVECDDINSTSTIKTVISLNLPTPLQEYVITQ---DLT
                                      |::||::: :|:  | ||:  :::   ||:
Inla    TNLTGLTLFNNQITDIDPLKNLTNLNRLELSSNTISDISALSGLTNLQQLSFGNQVTDLK
           90       100       110       120       130       140

           70        80        90       100       110       120
gp130   PLQNLTYMELYEKIYLTLSFFKNINKLTQLETIVTLSFNVTIPDDTIFPASLETFSIYKP
        ||:||| :|  :    ::| :: ::|||:||:::: : :::  :   : ::|:::|:
Inla    PLANLTTLERLDISSNKVSDISVLAKLTNLESLIATNNQISDITPLGILTNLDELSLNGN
          150       160       170       180       190       200

          130       140       150       160       170       180
gp130   SVPLSIAIFGSNIKNLYVNSPLTGYSIPTLINVNPYLENLQLPVTYYSGFPSNISLAFPN
         :  :|::::| ::|| ::  |::  |::|  ::   :  :| ::      :::|| :::
Inla    QLK-DIGTLAS-LTNL-TDLDLANNQISNLAPLSGLTKLTELKLG-----ANQIS-NISP
           210         220       230       240            250

          190       200       210       220       230       240
gp130   LQYLTIYVNNDMDQNNYHNFSISNIGVFKNLKGLDIEFTDSYNPQEFSINSFLSNVPVID
        |: ||  :| ::::|: ::  || |: :|||: |:: |::
Inla    LAGLTALTNLELNENQLED--ISPISNLKNLTYLTLYFNNISDISPVSSLTKLQRLFFYN
         260       270         280       290       300       310

          250       260       270       280       290       300
gp130   SLYIYGQGVTIDPSVGIIDLSYVKSKKFLSINIQESSLLNNCKGKSFKSPKKAYFRSNYN

Inla    NKVSDVSSLANLTNINWLSAGHNQISDLTPLANLTRITQLGLNDQAWTNAPVNYKANVSI
           320       330       340       350       360       370
```

Fig. 4. Sequence comparison of the membrane glycoprotein gp130 from *Dictyostelium discoideum* with internalin A from *Listeria* (Gaillard et al. 1991)

N-terminal sequencing of gp130, cDNA clones have been isolated from which a complete amino-acid sequence has been derived.

Gp130 is of interest for three reasons. First, its presence in growth-phase cells renders it possible that gp130 is involved in the attachment of bacteria to *D. discoideum* cells, and also in the formation of EDTA-sensitive contacts between growth-phase cells of *D. discoideum.* These contacts have been designated B contacts to distinguish them from csA-mediated adhesion (Beug et al. 1973). Antibody blocking experiments have pointed to a 126 kDa glycoprotein on the surface of growth-phase cells that may be responsible for contact-site B activity and for the interaction of *D. discoideum* cells with bacteria (Chadwick et al. 1984). Second, gp130 shows sequence relationships in its N-terminal domain not only to csA but also to internalin A (Fig. 4), a bacterial protein that is implicated in the invasion of epithelial cells by *Listeria* (Gaillard et al. 1991). Third, the sequence of the cloned gp130 is identical with that of "protein B" discovered by Yanagisawa and his colleagues during their analysis of sexual cell fusion between *D. discoideum* strains of different mating type (Fang et al. 1993). The authors found two transcripts encoding proteins with similar sequences that are differently regulated on the transcript level. Our results with an anti-peptide antibody against the N-terminal sequence also indicate two differently regulated proteins which are distinguishable by their electrophoretic mobilities. Together these results suggest that gp130 consists of two members of a family, with putative functions in cell adhesion and sexual cell fusion.

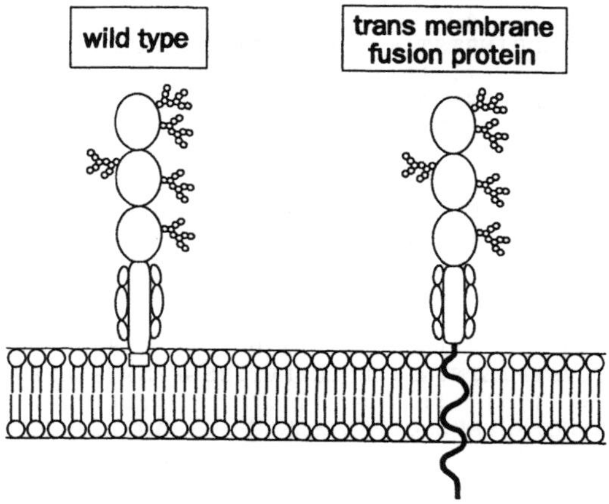

Fig. 5. Scheme of the PL-anchored csA glycoprotein *(left)* and of a fusion protein in which the lipid anchor is replaced by a transmembrane region *(right)*

4 Anchoring by Phospholipid Guarantees Long Persistence of the csA Protein on the Cell Surface

As in other phospholipid (PL) anchored proteins (Ferguson and Williams 1988), the csA translation product contains a sequence at its carboxy-terminal end which encompasses a signal for cleavage of the polypeptide chain and replacement of its hydrophobic C-terminus by the PL-anchor (Barth 1992). In order to examine whether the anchor is important for csA function, it has been replaced by the transmembrane region together with a short cytoplasmic tail from another cell-surface protein of *D. discoideum* (Fig. 5). The chimeric protein, which has retained the complete extracellular portion of the csA protein, enables growth-phase cells to form EDTA-stable cell contacts similar to the normal, PL-anchored csA. However, the turnover of the transmembrane version of the csA protein is much more rapid. The normal PL-anchored csA has an unusually long half-life of more than 6 h, the transmembrane form of less than 2 h. The fast turnover of the transmembrane version is due to its rapid internalization and degradation in lysosomes (Barth 1992). From these data it is inferred that the PL-anchor is less important for the function of the csA protein than for its prolonged persistence on the plasma membrane.

5 O-linked Oligosaccharide Residues Protect the csA Protein Against Protease

The O-glycosylated region of the csA protein connects the large N-terminal domain to the phospholipid anchor in the plasma membrane. O-linked oligosaccharides of csA, referred to as "type 2 carbohydrates" (Berthold et al. 1985) belong to a family of protein-linked oligosaccharide structures that are typical of *D. discoideum* cells. Monoclonal antibodies against these carbohydrates recognize in immunoblots differing sets of *Dictyostelium* proteins: some of them, such as mAb 19-14-19, label only two glycoproteins. Other antibodies label a large number, and each one another set of glycoproteins (Berthold et al. 1985). Thus, the structures of O-linked carbohydrates differ extensively, depending on the protein to which they are attached.

Mutants of *D. discoideum* defective in O-glycosylation can be obtained by antibody screening (Murray et al. 1984) and are easy to select with a fluorescence-activated cell sorter. Mutants in the *modB* locus lack all the epitopes of type 2 carbohydrates (Gerisch et al. 1985). These mutants show extremely weak EDTA-stable cell adhesion because the csA glycoprotein is proteolytically cleaved: an N-glycosylated 50-kDa fragment, representing the entire N-terminal domain of the csA molecule, is released into the medium (Hohmann et al. 1987a). Therefore, in order to determine whether or not O-linked oligosaccharides have a function in cell adhesion, a *modB* mutant has been mutagenized and a double mutant selected by cell sorting that strongly binds csA protein-specific antibody. The double mutant HG628 has the properties required for the adhesion assay: it contains normal amounts of csA on the cell surface although its defect in O-glycosylation is not reversed (Fig. 6). HG628 carries a second-site suppressor mutation in a gene responsible for the production of a cell-bound protease, which in *modB* mutants cleaves the csA protein along its proline-rich region. If subjected to the cell-adhesion assay in the presence of EDTA, HG628 behaves like wild type, which means that for its activity as a cell adhesion molecule the csA protein does not need to be decorated by O-linked carbohydrate chains. This result is noticeable since antibodies directed against the O-linked carbohydrate have been found to block csA-mediated cell adhesion (Loomis et al. 1985).

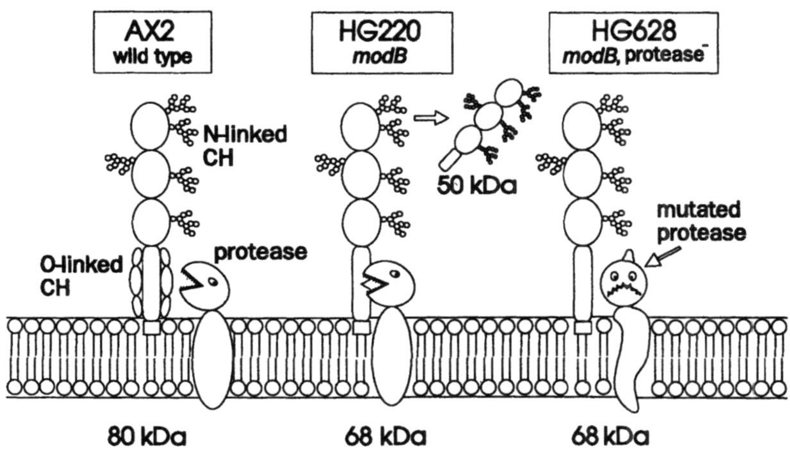

Fig. 6. Scheme of carbohydrate *(CH)* modifications of the csA-glycoprotein in the AX2 wild type and in *modB*-mutants lacking O-linked oligosaccharides. The normal, fully glycosylated csA has an apparent molecular mass of 80 kDa *(left)*. Without O-linked sugars the molecular mass shifts to 68 kDa and the protein is proteolytically cleaved in *D. discoideum* cells, giving rise to a soluble 50 kDa fragment *(center)*. Elimination of a protease by mutation results in a protein that is stable without O-linked sugars *(right)*

6 Puzzling Results on the Importance of N-Linked Carbohydrate Residues and Their Modifications

The sequence of the csA protein indicates a maximum of five N-glycosylation sites, and site-directed mutagenesis has shown that at least four of these sites (probably all of them) are used (Faix 1993). The N-linked oligosaccharide residues of the csA protein are distinguished by their high degree of sulfation (Stadler et al. 1983), and by their specific reactivity with an antibody, mAb 353, that requires fucose in the carbohydrate for binding. A role for fucose in oligosaccharides has been suggested from the release into the medium of an α-fucosidase by developing *D. discoideum* cells. This α-fucosidase has sequence relationships to human α-fucosidase (Müller-Taubenberger et al. 1989), and is coregulated with csA in cAMP-stimulated cells (Schopohl et al. 1992). The conspicuous features of N-linked oligosaccharides of the csA protein, and also the inhibition of cell adhesion by doses of tunicamycin low enough for synthesis and transport of the csA protein to proceed, have suggested a specific role of these carbohydrates and their modifications in csA function. In order to test this unambiguously, mutants have been employed that are either defective in the synthesis of L-fucose (Gonzales-Yanes et al. 1989) or show extremely low levels of sulfate incorporation (Knecht et al. 1984). In both types of mutants, EDTA-stable cell adhesiveness is not substantially altered. Since, according to these results, sulfation and fucosylation of the N-linked carbohydrates appear to be irrelevant for function of csA in cell adhesion, we have questioned whether these oligosaccharide residues are required at all. All five N-glycosylation sites have been inactivated by conversion of Asn into Gln residues in the acceptor consensus sequence Asn-X-Ser/Thr. The mutated protein has either been expressed in growth-phase cells of the wild type, or in aggregating cells of a csA-negative mutant. In both cases the mutated protein enables the cells to cohere to each other in the presence of EDTA.

Finally, in order to generate a csA protein that is neither N- nor O-glycosylated the mutated protein has been expressed in a *modB* mutant. Cells of the transformed *modB* strain still adhere to each other in the presence of EDTA, which indicates that neither one of the two types of carbohydrate chains is required for csA function. However, in the cells that produce csA devoid of both N- and O-linked carbohydrate, adhesion is weaker than predicted from the total amount of csA protein these cells express. This is in accord with immunofluorescence images showing most of this protein to accumulate within the cells, in structures near the nuclei which coincide with the location of the Golgi system (Fig. 7). In conclusion, modification by either N- or O-linked oligosaccharide residues is required for keeping the protein competent for transport through the Golgi to the plasma membrane.

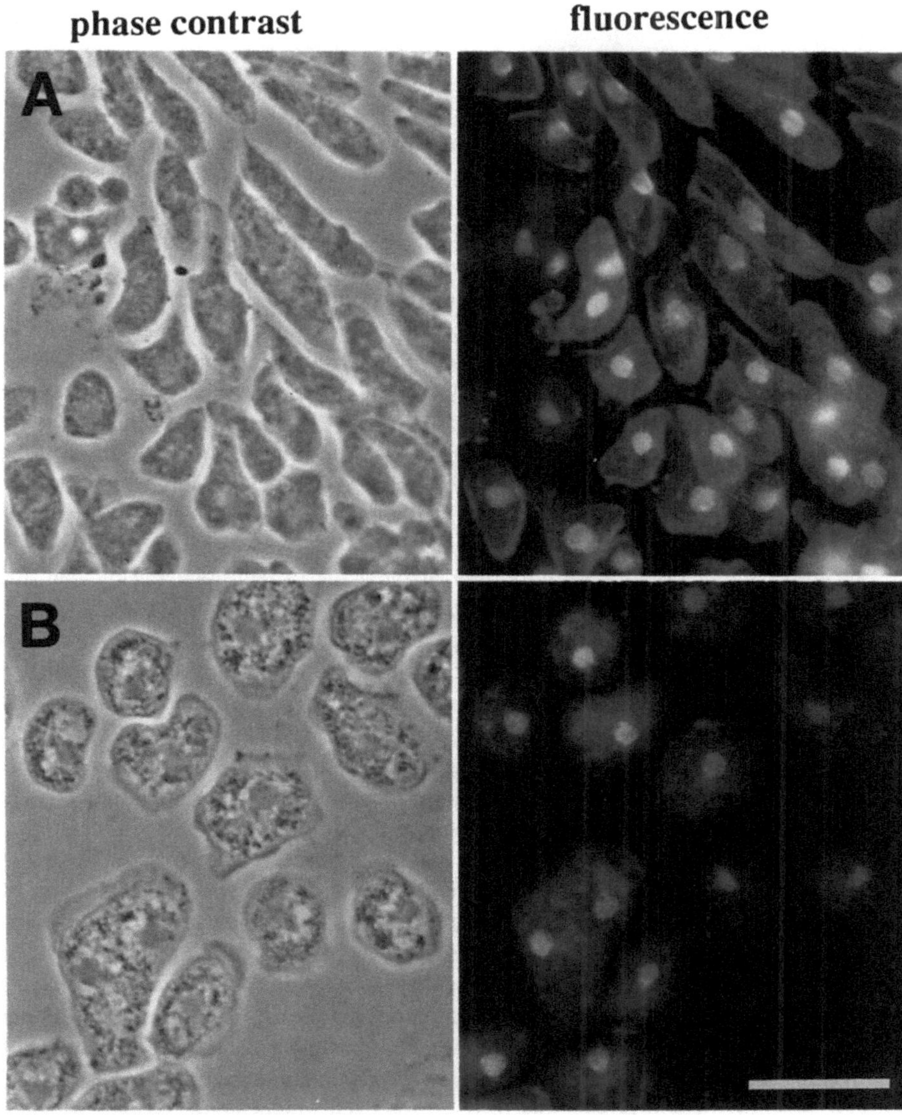

Fig. 7 A, B. Impairment of transport by elimination of N- and O-linked glycosylation. Fluorescent antibody labeling of the fully glycosylated csA protein is shown in aggregation-competent cells of wild-type AX2 (**A**). Labeling of csA protein that lacks N- and O-linked carbohydrates is demonstrated in growth-phase cells of HG628 (see Fig. 6) transformed with csA cDNA mutated to eliminate all five N-glycosylation sites (**B**). *Red color* indicates TRITC-antibody label of anti-csA mAb 71, *blue color* DAPI staining of nuclei. *Bar* 20 μm

7 Species Specificity of Cell Adhesion and a Role of Protein-Linked Carbohydrate in the Morphogenesis of *Polysphondylium*

Earlier work indicated that aggregating cells of *D. discoideum* distinguish "self" from "nonself" by their csA-dependent adhesion system (Bozzaro and Gerisch 1978). It should be pointed out here that the structure and the negative charges characteristic for the N-linked carbohydrate residues of the csA protein may play a role in making the cell adhesion selective. Species specificity of the cell adhesion has been assayed in these studies after allowing cells of *D. discoideum* and *Polysphondylium pallidum* to co-agglutinate in suspension and by demonstrating sorting-out within the cell clusters (Gerisch et al. 1980). This sorting-out has prompted us and other groups to investigate cell-membrane proteins and their glycosylation in *P. pallidum*. Although the two species belong to the same family, Dictyosteliaceae, of the Acrasiomycetes (Raper 1984), their carbohydrate structures are distinct, as indicated by the paucity of immunological cross-reactivities. Assays based on the blocking of cell adhesion by antibody Fab fragments have led to the isolation of a putative cell-adhesion protein of *P. pallidum* (Toda et al. 1984a), and to its sequencing at the cDNA level (Saito et al. 1993).

We wish to concentrate here on a case of blockage of cell adhesion by Fab that turned out to be misleading, but has nevertheless provided evidence for a role of protein-linked carbohydrates in morphogenesis. By screening of monoclonal antibodies for their adhesion inhibiting activity, one antibody, mAb 293, has been identified whose Fab fragments completely inhibit cell adhesion in *P. pallidum* without having an effect on adhesion in *D. discoideum* (Toda et al. 1984a). This antibody binds to an epitope that is common to many glycoproteins of *P. pallidum*. Free L-fucose inhibits this binding, indicating that the epitope recognized contains a fucose residue (Toda et al. 1984b). Mutant cells have been selected that do not bind mAb 293 (Francis et al. 1985). In the mutants obtained, absence of the epitope recognized by mAb 293 proved to be correlated with a delay in cell aggregation and with the formation of smaller aggregates. Nevertheless, it would be erroneous to infer from the adhesion blocking effect of the antibody and from the impaired aggregation in the mutants a role of carbohydrate structure in the adhesion of aggregating cells. A detailed analysis has shown that the fucosylation which is required for mAb 293 binding is under developmental control. The fucose-containing epitope appears early in cell cultures of *P. pallidum*, already during exponential growth. The mutants selected by use of the antibody are impaired in proceeding through the early stages of development at the normal rate. This has been shown by delayed expression of a marker protein of development, a lectin called pallidin. Obviously, a master gene for the control of development is affected in the mutants. Impaired cell aggregation, delayed expression of pallidin, and retarded generation of the epitope recognized by mAb 293 are pleiotropic effects of a mutation in this master gene. There is nothing in these results, therefore, that substantiates the believe that the fucose-containing carbohydrate structure is required for cell aggregation.

A second set of mutants has been selected by employing mAb 293 with one change in the protocol: the mutated cells have not been sorted immediately after the end of the growth phase, as before, but after 20 h or more of starvation. This has the

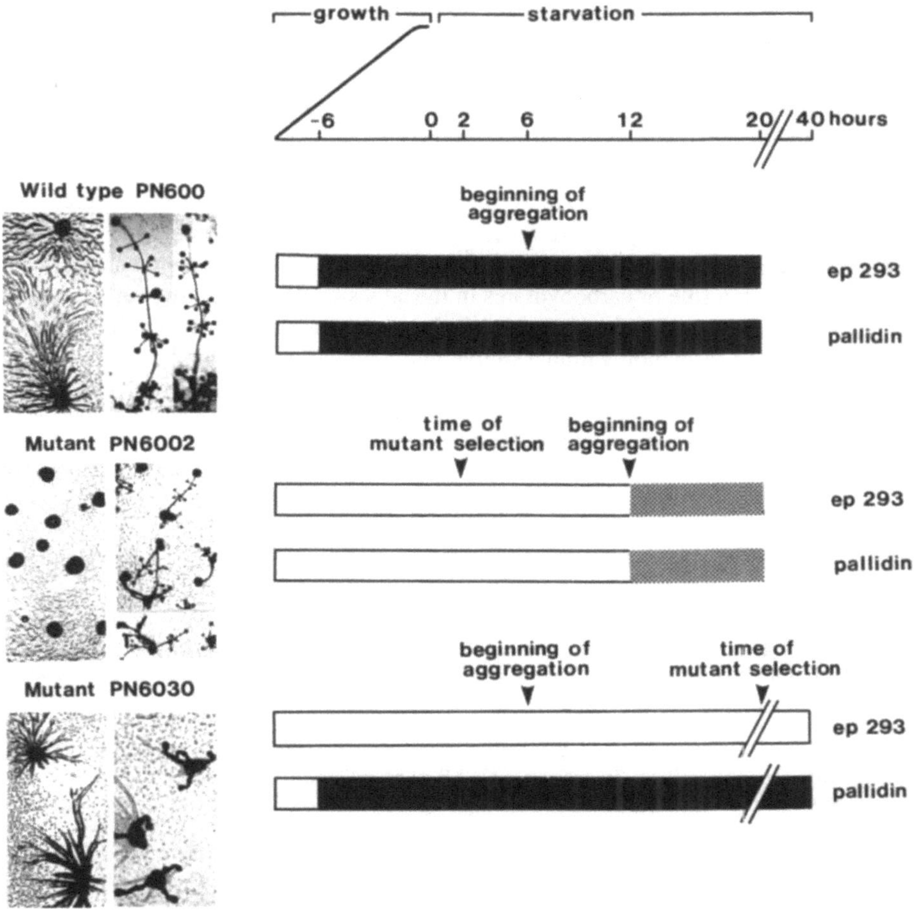

Fig. 8. Protein-linked carbohydrate modification and morphogenesis in *Polysphondylium pallidum* wild type and mutants. *On the left* aggregation patterns and final stages of development are shown for wild type PN600, for mutant PN6002 characterized by delayed expression of the carbohydrate epitope ep 293, and for mutant PN6030 representing mutants that are completely defective in ep 293 expression. *On the right* block diagrams illustrate expression of the L-fucose containing epitope ep 293 on glycoproteins and of the developmental marker protein pallidin during growth and starvation (*open* no expression; *shaded* weak expression; *filled* strong expression). In the wild type *(top)*, both ep 293 and pallidin are expressed at an early stage of exponential growth. In the pleiotropic mutant PN6002 showing delayed and incomplete expression of both markers *(center)*, aggregation is characterized by impairment of stream formation of cells moving towards aggregation centers. In PN6030 completely defective in ep 293 expression *(bottom)*, the aggregation pattern is normal, but fruiting body formation is severely impaired

advantage that no mutants are selected which express the epitope after a delay. Accordingly, the three mutants obtained never bind mAb 293 during their development, but express the marker pallidin at the same early stage as wild type (Fig. 8). The phenotype of these mutants is new and unpredicted. Cell aggregation is normal, which demonstrates that the epitope recognized by mAb 293 is not required for the interaction of aggregating cells. The blockage of cell adhesion by Fab of this antibody may simply indicate that one of the proteins carrying this epitope is important for adhesion. The most interesting point is that morphogenesis at the final stage of development is dramatically disturbed in all three mutants that lack the epitope. *P. pallidum* wild type forms elegant, regulary branched fruiting bodies. Only irregularly branched, rudimentary structures are formed by the mutants (Fig. 8). In these structures the mutant cells are covered by cell walls as in the terminal stage of wild-type development, but spores and stalk cells are not distinguishable. Thus, the approach that has failed to establish a role of carbohydrates in the adhesion of aggregating cells has finally provided evidence for the importance of carbohydrate structure in cell differentiation and pattern formation within a simple multicellular organism.

Acknowledgments. This work was supported by the Sonderforschungsbereich 266 of the Deutsche Forschungsgemeinschaft. We thank Christina Heizer for immunolabelling.

References

Barth A (1992) Untersuchungen zur Funktion des Phospholipid-Ankers eines Zelladhäsionsproteins von *Dictyostelium discoideum* RAPER. Ph D Thesis: Univ Hamburg

Bertholdt G, Stadler J, Bozzaro S, Fichtner B, Gerisch G (1985) Carbohydrate and other epitopes of the contact site A glycoprotein of *Dictyostelium discoideum* as characterized by monoclonal antibodies. Cell Diff 16:187–202

Beug H, Katz FE, Gerisch G (1973) Dynamics of antigenic membrane sites relating to cell aggregation in *Dictyostelium discoideum.* J Cell Biol 56:647–658

Bozzaro S, Gerisch G (1978) Contact sites in aggregating cells of *Polysphondylium pallidum.* J Mol Biol 120:265–279

Bozzaro S, Merkl R, Gerisch G (1987) Cell adhesion: Its quantification, assay of the molecules involved, and selection of defective mutants in *Dictyostelium* and *Polysphondylium.* Methods Cell Biol 28:359–385.

Chadwick CM, Ellison JE, Garrod DR (1984) Dual role for *Dictyostelium* contact site B in phagocytosis and developmental size regulation. Nature 307:646–647

Desbarats L, Lam TY, Wong LM, Siu C-H (1992) Identification of a unique cAMP-response element in the gene encoding the cell adhesion molecule gp80 in *Dictyostelium discoideum.* J Biol Chem 267:19655–19664

Faix J, Gerisch G, Noegel AA (1990) Constitutive overexpression of the contact site A glycoprotein enables growth-phase cells of *Dictyostelium discoideum* to aggregate. EMBO J 9:2709–2716

Faix J, Gerisch G, Noegel AA (1992) Overexpression of the csA cell adhesion molecule under its own cAMP-regulated promoter impairs morphogenesis in *Dictyostelium.* J Cell Sci 102:203–214

Faix J (1993) Gezielte Expression und Mutagenese des csA-Glycoproteins von *Dictyostelium discoideum* zur Untersuchung der strukturellen Voraussetzungen seiner Funktion als Zelladhäsionsmolekül. Ph D Thesis, Univ Regensburg

Fang H, Higa M, Suzuki K, Aiba K, Urushihara H, Yanagisawa K (1993) Molecular cloning and characterization of two genes encoding gp138, a cell surface glycoprotein involved in the sexual cell fusion of *Dictyostelium discoideum.* Dev Biol 156:201–208

Ferguson MAJ, Williams AF (1988) Cell-surface anchoring of proteins via glycosylphosphatidyl-inositol structures. Annu Rev Biochem 57:285–320

Francis D, Toda K, Merkl R, Hatfield T, Gerisch G (1985) Mutants of *Polysphondylium pallidum* altered in cell aggregation and in the expression of a carbohydrate epitope on cell surface glycoproteins. EMBO J 4:2525–2532

Gaillard J-L, Berche P, Frehel C, Gouin E, Cossart P (1991) Entry of L. monocytogenes into cells is mediated by internalin, a repeat protein reminiscent of surface antigens from gram-positive cocci. Cell 65:1127–1141

Gerisch G, Krelle H, Bozzaro S, Eitle E, Guggenheim R (1980) Analysis of cell adhesion in *Dictyostelium* and *Polysphondylium* by the use of *Fab*. In: Curtis ASG, Pitts JD (eds) Cell adhesion and motility. University Press, Cambridge, pp 293–307

Gerisch G, Weinhart U, Bertholdt G, Claviez M, Stadler J (1985) Incomplete contact site A glycoprotein in HL220, a modB mutant of *Dictyostelium discoideum*. J Cell Sci 73:49–68.

Gonzales-Yanes B, Mandell RB, Girard M, Henry S, Aparicio O, Gritzali M, Brown RD Jr, Erdos GW, West CM (1989) The spore coat of a fucosylation mutant in *Dictyostelium discoideum*. Dev Biol 133:576–587

Hohmann H-P, Gerisch G, Lee RWH, Huttner WB (1985) Cell-free sulfation of the contact site A glycoprotein of *Dictyostelium discoideum* and of a partially glycosylated precursor. J Biol Chem 260:13869–13878

Hohmann H-P, Bozzaro S, Merkl R, Wallraff E, Yoshida M, Weinhart U, Gerisch G (1987a) Post-translational glycosylation of the contact site A protein of *Dictyostelium discoideum* is important for stability but not for its function in cell adhesion. EMBO J 6:3663–3671

Hohmann H-P, Bozzaro S, Yoshida M, Merkl R, Gerisch G (1987b) Two-step glycosylation of the contact site A protein of *Dictyostelium discoideum* and transport of an incompletely glycosylated form to the cell surface. J Biol Chem 262:16618–16624

Knecht DA, Dimond RL, Wheeler S, Loomis WF (1984) Antigenic determinants shared by lysosomal proteins of *Dictyostelium discoideum*. J Biol Chem 259:10633–10640

Loomis WF, Wheeler SA, Springer WR, Barondes SH (1985) Adhesion mutants of *Dictyostelium discoideum* lacking the saccharide determinant recognized by two adhesion-blocking monoclonal antibodies. Dev Biol 109:111–117

Müller K, Gerisch G (1978) A specific glycoprotein as the target site of adhesion blocking Fab in aggregating *Dictyostelium* cells. Nature 274:445–449

Müller-Taubenberger A (1989) Sequenzierung und funktionelle Charakterisierung von entwicklungsregulierten Genen in *Dictyostelium discoideum*. Ph D Thesis, Univ Stuttgart

Müller-Taubenberger A, Westphal M, Noegel A, Gerisch G (1989) A developmentally regulated gene product from *Dictyostelium discoideum* shows high homology to human α-L-fucosidase. FEBS Lett 246:185–192

Murray BA, Yee LD, Loomis WF (1981) Immunological analysis of a glycoprotein (contact sites A) involved in intercellular adhesion of *Dictyostelium discoideum*. J Supramol Struct Cell Biochem 17:197–211

Murray BA, Wheeler S, Jongens T, Loomis WF (1984) Mutations affecting a surface glycoprotein, gp80, of *Dictyostelium discoideum*. Mol Cell Biol 4:514–519

Noegel A, Harloff C, Hirth P, Merkl R, Modersitzki M, Stadler J, Weinhart U, Westphal M, Gerisch G (1985) Probing an adhesion mutant of *Dictyostelium discoideum* with cDNA clones and monoclonal antibodies indicates a specific defect in the contact site A glycoprotein. EMBO J 4:3805–3810

Noegel A, Gerisch G, Stadler J, Westphal M (1986) Complete sequence and transcript regulation of a cell adhesion protein from aggregating *Dictyostelium* cells. EMBO J 5:1473–1476

Raper KB (1984) The Dictyostelids. University Press, Princeton, p 227

Saito T, Kumazaki T, Ochiai H (1993) A purification method and N-glycosylation sites of a 36-cysteine-containing, putative cell/cell adhesion glycoprotein gp64 of the cellular slime mold, *Polysphondylium pallidum*. Eur J Biochem 211:147–155

Saxe III CL, Klein P, Sun TJ, Kimmel AR, Devreotes PN (1988) Structure and expression of the cAMP cell-surface receptor. Dev Gen 9:227–235

Schopohl D, Müller-Taubenberger A, Orthen B, Hess H, Reutter W (1992) Purification and properties of a secreted and developmentally regulated α-L-fucosidase from *Dictyostelium discoideum*. J Biol Chem 267:2400–2405

Stadler J, Gerisch G, Bauer G, Suchanek C, Huttner WB (1983) In vivo sulfation of the contact site A glycoprotein of *Dictyostelium discoideum*. EMBO J 2:1137–1143

Stadler J, Keenan TW, Bauer G, Gerisch G (1989) The contact site A glycoprotein of *Dictyostelium discoideum* carries a phospholipid anchor of a novel type. EMBO J 8:371–377

Toda K, Bozzaro S, Lottspeich F, Merkl R, Gerisch G (1984a) Monoclonal anti-glycoprotein antibody that blocks cell adhesion in *Polysphondylium pallidum*. Eur J Biochem 140:73–81

Toda K, Tharanathan RN, Bozzaro S, Gerisch G (1984b) Monoclonal antibodies that block cell adhesion in *Polysphondylium pallidum*: reaction with L-fucose, a terminal sugar in cell-surface glycoproteins. Eur J Biochem 143:477–481

Towards Understanding Roles of Oligosaccharides as Recognition Structures

T. Feizi[1]

1 Introduction

Understanding the biological roles of the diverse oligosaccharides of glycoproteins and glycolipids has been a major challenge in cell biology. From work with monoclonal antibodies, it emerged that there are remarkable spatio-temporal patternings in the display of several oligosaccharide antigens at the surface of cells and in extracellular matrices during stages of embryonic development and cellular differentiation. There is a different repertoire of oligosaccharide antigens in various differentiated cells and tissues, and there are predictable changes in these antigens in malignancy (Feizi 1985). This raised the possibility that such oligosaccharides may be recognition structures for proteins (endogenous lectins) which determine the way cells migrate or respond to various micro-environments (Feizi 1981; 1989; Feizi and Childs 1987; Loveless et al. 1990). However, rapid progress in elucidating the roles of specific oligosaccharides (particulary those associated with glycoproteins) as recognition elements has been hampered, in part, by the lack of straightforward biological assay methods. To meet this challenge, we introduced a technology which enables oligosaccharide probes (neoglycolipids) to be generated from mixtures of oligosaccharides released from glycoproteins, and from desired structurally defined or chemically synthesised oligosaccharides (Tang et al. 1985; Stoll et al. 1988; 1990; Feizi et al. 1993). This has proven to be not only a powerful means of singling out, even from heterogenous mixtures, the ligands for endogenous carbohydrate-binding proteins (lectins), but in addition it constitutes a microsequencing strategy for oligosaccharides, since the neoglycolipids have unique ionization properties in mass spectometry (Lawson et al. 1990; Feizi et al. 1993). Thus a picture is emerging of the way in which biological specificities can arise even through recognition of commonly occurring oligosaccharides by endogenous carbohydrate-binding proteins. In this chapter are discussed (1) the way in which the monoclonal antibody approach served to identify certain oligosaccharide structures of the blood group family which behave as differentiation antigens of human leukocytes or of cancer cells, (2) the emergence of certain members of this family of oligosaccharides as ligands for cell adhesion molecules (the selectins) which have crucial roles at the initial stages of leukocyte recruitment in inflammation, and (3), the way in which the neoglycolipid technology is helping with the identification of novel ligands for the selectins.

[1] Glycoconjugates Section, MRC Clinical Research Centre, Harrow, Middlesex HA1 3UJ, UK.

44. Colloquium Mosbach 1993
Glyco- and Cellbiology

2 Hybridoma Antibodies, and the Emergence of Cell Surface Saccharides as Differentiation Antigens of Human Leukocytes

With the advent of the hybridoma technology (Kohler and Milstein 1975), antibodies could be readily raised to surface markers that distinguish embryonic cells from those of the adult, or various differentiated cells from one another, or malignant cells from their normal counterparts. The hope that such antibodies would reveal unique surface markers of embryonic stage, differentiated state or neoplastic state was replaced by the realization that such antigenic "markers" are mainly carbohydrate structures of glycoproteins and glycolipids (Feizi 1985). For example, the stage-specific embryonic antigen (designated SSEA-1) of the mouse embryo was identified (Gooi et al. 1981) as α1–3 fucosylated type 2 blood group chains, also known as Le^x:

```
Galβ1–4GlcNAc                     SSEA–1; Le^x
        | 1,3
        Fucα
```

This antigen appears coincident with the first cell-to-cell adhesion event (compaction) of the eight-cell embryo (Solter and Knowles 1978). This assignment may have been an early clue to the Le^x structure being a ligand in cell adhesion and migration events (Gooi et al. 1981; Feizi 1981). Although such a cognate protein in embryonic cell adhesion has not yet been identified, it has been shown that the Le^x structure can support adhesion of the endothelial adhesion molecule E-selectin, as elaborated in Section 4.1.

In later studies, the Le^x structure was identified as a distinctive marker of human granulocytes among blood cells (Feizi 1983; Gooi et al. 1983). Numerous antibodies raised to granulocytes are directed to this and related carbohydrate structures (Feizi 1984; Magnani 1984) which are now collectively termed CD15. Using sequence-specific monoclonal antibodies, it was found that sialyl-Le^x is expressed on human granulocytes and monocytes, and thus could distinguish these cells from the majority of other types of blood cells (Thorpe and Feizi 1984).

```
Galβ1–4GlcNAc                     sialyl–Le^x
|       | 1,3
NeuAcα Fucα
```

A further antigen common to granulocytes and monocytes recognized by a hybridoma antibody, VIM-2, was identified as a long chain sialo-fuco-oligosaccharide in this series with fucosylation at an internal rather than an outer N-acetylglucosamine (Uemura et al. 1985; Macher et al. 1988); this antigen is now designated CD65.

```
Galβ1–4GlcNAcβ1–3Galβ1–4GlcNAcβ1–3Galβ–   VIM–2; CD65
| 2,3                     | 1,3
NeuAcα                    Fucα
```

Were these carbohydrate antigens "area codes" that determine cell migration patterns, and were they perhaps ligands in macromolecular interactions in health and disease? (Feizi 1981, 1985, 1989; Feizi and Childs 1987). Until recently, scepticism prevailed among biologists that oligosaccharide antigens could have such important roles. One reason was that the patterns of expression of these antigens are not conserved among vertebrates; for example, in the mouse, Le^x behaves as an embryonic antigen and is barely detectable in the adult animal. By contrast, in the human, this antigen has quite a wide distribution in the body, although among blood cells it is highly restricted to myeloid cells (reviewed by Gooi et al. 1983). However, during the last 3 years, a dramatic change has occurred in attitudes to the biology of carbohydrates. With developments in the molecular cloning and sequencing of endothelial cell adhesion molecules (selectins) that bind granulocytes and monocytes, and the finding of lectin domains on these proteins, international research has been focused on the Le^x, sialyl-Le^x, and VIM-2 antigens of leukocytes as obvious candidate ligands. As discussed below, roles have been established for specific oligosaccharides in this series as ligands in leukocyte-endothelial interactions that promote leukocyte recruitment in inflammation.

3 Selectins, and the Emergence of Cell Surface Oligosaccharides as Ligands in Leukocyte-Endothelial Interactions

The selectins (E-selectin, P-selectin, and L-selectin) are integral membrane proteins, each with an amino terminal lectin-like domain which is in tandem with an epidermal growth factor-like, and multiple complement regulatory-like domains as depicted schematically in Fig. 1 (for reviews see Harlan and Liu 1992). E-selectin is expressed on cytokine-stimulated endothelia; P-selectin becomes exteriorized from intracellular granules onto the surface of endothelial cells and platelets, under the influence of thrombin and other cell activators; while L-selectin is constitutively expressed on subsets of all types of leukocytes. It is now widely accepted that the sectins have key roles at the initial stages of leukocyte recruitment in inflammation, for example in leukocyte entrapment in post-capillary venules in inflammation (E- and P-selectins), or in lymphocyte homing to peripheral lymph nodes via high endothelial venules (L-selectin). The rolling phenomenon that leukocytes display along the endothelium prior to leukocyte arrest in the circulation is mediated by selectins, and is a prerequisite for the tight leukocyte adhesion and transmigration events that follow and are mediated by integrins and their counter-receptors. If the selectin-mediated rolling events are inhibited, then the subsequent adhesion events cannot proceed. For this reason, the selectins and their carbohydrate ligands are attracting enormous interest, and a whole new field, based on knowledge of selectin-mediated interactions, has become a growth area and the subject of vigorous research in medical biology and biotechnology targeted at the design of adhesion inhibitors that may serve as drugs in the management of disorders of inflammation (Hodgson 1992; Edgington 1992).

Different approaches have been made by various groups in the search for selectin ligands (for reviews see Brandley et al. 1990; Feizi 1991a, b; Harland and Liu 1992). By inhibition with monoclonal antibodies and by transfection experiments (transfection of fucosyltransferases into cells lacking these enzymes), it has been clearly shown that 3'-sialyl-Le^x is a ligand for E-selectin. For example, Lowe et al. (1990) transfected cDNA for an α1–3/4 fucosyltransferase into cells that normally lack this

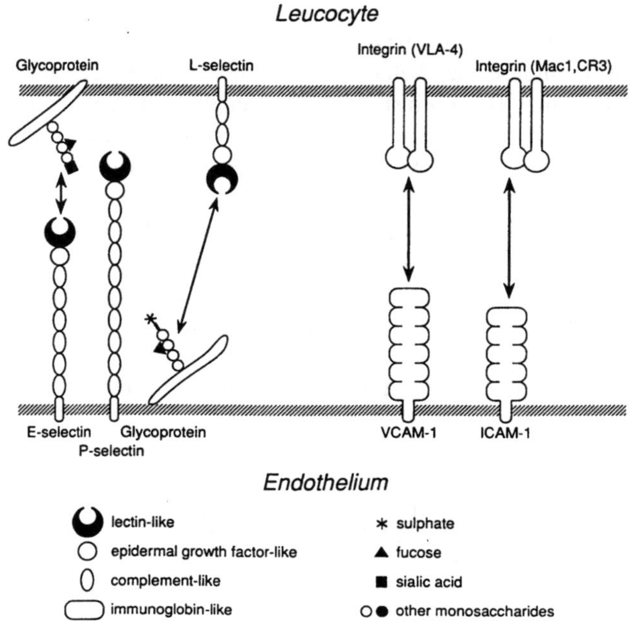

Fig. 1. Schematic representation of the surface of a leukocyte and of an endothelial cell depicting *(left)* the three selectin molecules, E-, P-, and L-, and their oligosaccharide ligands displayed on glycoproteins which mediate the initial leukocyte-endothelial interactions in inflammation; these interactions give rise to leukocyte rolling along the endothelial surface, an essential first step in leukocyte recruitment in inflammation and a prerequisite for subsequent tighter adhesive events that are mediated by the integrins and their counter-receptors *(right)*

enzyme and showed that E-selectin ligands can be elicited by this means. This enzyme fucosylates effectively the 3'-sialyl sequence NeuAcα2–3Galβ1-4GlcNAc- to form the sialyl-Le^x structure. There have been conflicting reports, however, on the ability of a second fucosyltransferase, termed ELFT, to elicit E-selectin ligands (for reviews see Feizi 1991b; Larkin et al. 1992). This fucosyltransferase cannot effectively fucosylate the penultimate N-acetylglucosamine of the sialyl backbones. It readily fucosylates the asialo backbone (to form the Le^x antigen) or the internal N-acetylglucosamine of an extended backbone to form the VIM-2 antigen structure. Thus, powerful though the transfection experiments are, precise structural assignments could not be made by this means. Moreover, the cell transfection/cloning experiments may well be complicated by clone-to-clone variations. For example, Childs et al. (1980) have noted remarkable cell-to-cell variations in the glycosylation patterns within established cell lines. There has, therefore, been a need for direct binding studies (single parameter analyses) using structurally characterized oligosaccharides to determine precisely which oligosaccharides constitute ligands for E-selectin, for example by the neoglycolipid technology as elaborated below.

As for human P-selectin, the information on the carbohydrate ligands has been particularly conflicting (reviewed by Feizi 1991a, b). A possible complicating factor is the large size of the extracellular domain and the numerous complement regulatory-like domains which may mediate adhesion events in their own right. A glycoprotein approximately 120 kDa, to which P-selectin binds, has been identified on human myeloid cells (Moore et al. 1992).

Most studies on L-selectin ligands have been performed with the murine homologue. Distinct glycoproteins to which this adhesion molecule binds have been described. These are 50-kDa and 90-kDa glycoproteins containing sulfate, fucose, and sialic acid from murine peripheral lymph nodes (Imai et al. 1991; Lasky et al. 1992; Imai et al. 1993) and a major 90-105-kDa glycoprotein from human tonsils (Berg et al. 1991). The oligosaccharide ligands on these glycoproteins have not been characterized, although in separate studies L-selectin was found to bind to a glycolipid containing the 3'-sialyl-Le^x-active hexasaccharide (Foxall et al. 1992).

4 Contributions by the Neoglycolipid Approach

4.1 Studies of E-Selectin Binding Specificity Using Neoglycolipids Derived from a Series of Structurally Defined Oligosaccharides

Our group have been making two approaches to elucidating the range of oligosaccharides recognized by the selectins. The first approach has been to evaluate, in *direct* binding experiments, the reactivity of E-selectin towards an array of structurally defined oligosaccharides that have been chemically linked to lipid (neoglycolipids), and resolved on chromatograms or immobilized on plastic microwells in the presence of carrier lipids (Larkin et al. 1992). These have included oligosaccharides isolated from human milk which encompass the blood group antigens and related carbohydrate differentiation antigens (Table 1). Two reference compounds, 3'-sialyl-Le^x and 6'-sialyl-Le^x glycosphingolipids, were provided as chemically synthesized compounds by Professors Akira Hasegawa and Makoto Kiso (Gifu, Japan). In the second approach (discussed in Section 4.2), E-selectin ligands have been sought among unknown oligosaccharides released from a mucin-type glycoprotein.

In exploratory experiments, we found that the soluble form of E-selectin gave no detectable binding, while Chinese hamster ovary (CHO) cells transfected to express a high level of E-selectin (as assessed by immunocytofluorimetry), bound not only to the 3'-sialyl-Le^x glycolipid, and to the 3'-sialyl-Le^x-active neoglycolipid (3'-S-LNFP-III) but also to the linkage isomer 3'-sialyl-Le^a (3'-S-LNFP-II). The specificity of binding was shown by a lack of binding to (1) the 6'-sialyl analogue, 6'-S'LNFP-III, (2) analogues lacking sialic acid and fucose, (3) several neoglycolipids derived from N-glycosylated proteins that are bound by conglutinin and the mannan-binding proteins (Mizuochi et al. 1989; Childs et al. 1989), and (4) glycolipids bound by the pulmonary surfactant protein A (Childs et al. 1992) Two other laboratories have also observed binding to the 3'-sialyl-Le^a sequence which behaves as a tumor-associated antigen in certain adenocarcinomas (Berg et al. 1991; Takada et al. 1991). The nonsialylated Le^x and Le^a sequences, LNFP-III and LNFP-II also support binding

Table 1. CH0-E-selectin cell binding to lipid linked oligosaccharides chromatographed on silica gel plates. The oligosaccharides were tested as neoglycolipids after conjugation to the lipid DPPE with the exception of 3'S-LNFP-III and 6'S-LNFP-III, which were tested as chemically synthesized glycosphingolipids. The binding results fell into three categories shown as –, +, ++, denoting no binding, moderately strong binding (comparable to that observed with LNFP-II and LNFP-III) and very strong binding, respectively, at the highest level tested (approx. 1 nmol)

Designation	Structure	Binding
Linear oligosaccharides with Type 1 termini		
LNT	Galβ1–3GlcNAcβ1–3Galβ1–4Glc	–
LNFP-I	Galβ1–3GlcNAcβ1–3Galβ1–4Glc | 1,2 Fucα	–
LNFP-II	Galβ1–3GlcNAcβ1–3Galβ1–4Glc | 1,4 (on GlcNAc) Fucα	+
LNDFH-I	Galβ1–3GlcNAcβ1–3Galβ1–4Glc | 1,2 | 1,4 Fucα Fucα	+
A-hepta	GalNAcα1–3Galβ1–3GlcNAcβ1–3Galβ1–Glc | 1,2 | 1,4 Fucα Fucα	–
B-hepta	Galα1–3Galβ1–3GlcNAcβ1–3Galβ1–Glc | 1,2 | 1,4 Fucα Fucα	–
3'S-LNFP-II	Galβ1–3GlcNAcβ1–3Galβ1–4Glc | 2,3 | 1,4 NeuAcα Fucα	++
3'S-LNT	Galβ1–3GlcNAcβ1–3Galβ1–4Glc | 2,3 NeuAcα	–
6'S-LNT	Galβ1–3GlcNAcβ1–3Galβ1–4Glc | 2,6 (on GlcNAc) NeuAcα	–
3'S-6'S LNT	Galβ1–3GlcNAcβ1–3Galβ1–4Glc | 2,3 | 2,6 NeuAcα NeuAcα	–
FpLNH	Galβ1–3GlcNAcβ1–3Galβ1–4GlcNAcβ1–3Galβ1–4Glc | 1,3 (on second GlcNAc) Fucα	–
DFpLNH	Galβ1–3GlcNAcβ1–3Galβ1–4GlcNAcβ1–3Galβ1–4Glc | 1,4 | 1,3 Fucα Fucα	+

Table 1 (continued)

Designation	Structure	Binding
Linear oligosaccharides with Type 2 termini		
3'FL	Galβ1–4Glc | 1,3 Fucα	–
3'FLN	Galβ1–4GlcNAc | 1,3 Fucα	–
3'S–3'FL	Galβ1–4Glc | 2,3 | 1,3 NeuAcα Fucα	–
LNNT	Galβ1–4GlcNAcβ1–3Galβ1–4Glc	–
LNFP-III	Galβ1–4GlcNAcβ1–3Galβ1–4Glc | 1,3 Fucα	+
6'S-LNNT	Galβ1–4GlcNAcβ1–3Galβ1–4Glc | 2,6 NeuAcα	–
3'S-LNFP-III	Galβ1–4GlcNAcβ1–3Galβ1–4Glc | 2,3 | 1,3 NeuAcα Fucα	++
6'S-LNFP-III	Galβ1–4GlcNAcβ1–3Galβ1–4Glc | 2,6 | 1,3 NeuAcα Fucα	–

(Table 1). A tenfold greater amount of LNFP-II, and an even greater amount of LNFP-III are required to give binding equivalent to that of the sialyl analogues.

The blood group H-active oligosaccharide (LNFP-I) was not bound by E-selectin (Table 1). However, the presence of the blood group H fucose does not hinder binding when associated with the Le^a structure, for the Le^b-active oligosaccharide, LNDFH-I, supported binding. The additional presence of the blood group A monosaccharide N-acetylgalactosamine α1–3 as in A heptasaccharide (Table 1), or the blood group B monosaccharide Galα1–3 (unpubl.) inhibited E-selectin binding.

We have evidence that the presence of fucose linked to the subterminal rather than to an internal N-acetylglucosamine is required for E-selectin binding: there was no binding to FpLNH, while DFpLNH was bound, and the strength of binding was equal to that of LNFP-II. Thus, our results cast doubt on the involvement of the VIM-2 oligosaccharide sequence as a ligand for E-selectin, although binding studies with the 3'-sialyl analogue of DFpLNH will be required before ruling out such an assignment.

We have observed that E-selectin binding to the clustered lipid-linked oligosaccharide ligands is highly dependent on the density of surface expression of the membrane-associated adhesion molecule: binding experiments with transfected CHO cells that express different levels of E-selectin (as assessed by immunocytofluorimetry) have shown that there is a threshold density of E-selectin required for binding of the sialyl sequences, and binding to the nonsialyl sequences is a property only of cells expressing the highest density of E-selectin (Larkin et al. 1992).

Our results provide a possible explanation for the apparently conflicting results of in vitro E-selectin binding experiments using CHO cells transfected with α1–3 fucosylstransferase, ELFT. We predict that E-selectin binding would be elicited given (1) a sufficiently dense cell surface presentation of the nonsialylated products of this enzyme with fucose 3-linked to the outer N-acetylglucosamine, and (2) a sufficiently dense cell surface expression of E-selectin.

As discussed earlier, the Le^x and Le^a series of oligosaccharides are rather widely distributed in the body, for example, in many epithelia and their secretions (Gooi et al. 1983). Despite this, biological specificity in the blood vascular compartment is mediated through the regulated, high density of E-selectin expression only at desired sites on the endothelium in inflammation, coupled with the high levels of expression of the fuco-oligosaccharide ligands on granulocytes and monocytes.

Our findings may also have an important bearing on malignant disease. There is large body of evidence that Le^x, Le^a and the sialyl analogues, as well as blood groups Le^a, Le^b, ALe^b and BLe^b, are variably expressed at the surface of adenocarcinomas. A loss of epithelial cell polarity results in an extremely high density of expression of the antigens pericellularly. Some of the epithelial tumors also secrete glycoproteins that carry these antigens. Adhesion of human carcinoma cells to cytokine-stimulated endothelial cells has been documented previously by others. We have suggested that tumor cells strongly expressing E-selectin ligands at their surface may "highjack" the leucocyte-endothelial adhesion machinery and give rise to metastases, and that the metastatic potential of epithelial tumors in vivo may additionally be dependent on the amounts of soluble glycoproteins with E-selectin inhibitory activities that they release (Larkin et al. 1992). Thus, it will be important to investigate in detail the influence of the blood group AB0 status of individual patients on the display of E-selectin ligands on soluble and membrane-associated glycoconjugates produced by carcinomas of the distal colon. For, in contrast to the normal distal colon epithelium, which characteristically lacks these blood group antigens, the adenocarcinomas in this region of the colon express the blood group A, B, H, and Le^b antigens as "neo-antigens".

4.2 Novel E-Selectin Ligands Found on an Ovarian Cystadenoma Glycoprotein

In a second investigation of the adhesive specificity of E-selectin, we generated a "library" of neoglycolipids from the O-linked oligosaccharides on a human ovarian cystadenoma glycoprotein to which E-selectin binds strongly. Oligosaccharides were released from the glycoprotein by a mild β-elimination procedure. Numerous E-selectin-binding oligosaccharides were detected among both the acidic and non-acidic oligosaccharide fractions. So far, we have characterized the fastest migrating (smallest) component from one of several acidic oligosaccharide fractions (Fig. 2). Others are under investigation. By fast atom bombardment mass spectrometric

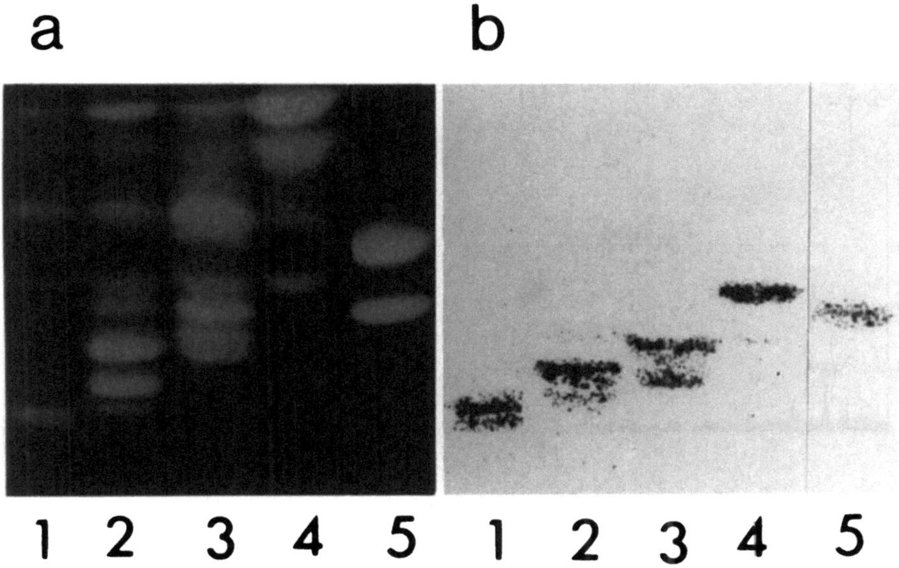

Fig. 2 a, b. Chromatogram overlay assay showing E-selectin binding to neoglycolipids chromatographed on silica gel plates. Neoglycolipids were revealed chemically with primulin (**a**) or overlaid with ^{3}H-labeled CH0 E-selectin cells and binding detected by fluorography (**b**). *Lanes 1–4* contained mixtures of neoglycolipids made from an acidic oligosaccharide fraction obtained from an ovarian cystadenoma glycoprotein and fractionated by size; *lane 5* contained neoglycolipid standards derived from lacto-N-tetraose, LNT, *(upper band)* which is not bound by E-selectin and LNFP-II *(lower band)*, 2 nmol of each. The fastest migrating component *(lane 4)* bound by E-selectin was investigated further. (Yuen et al. 1992)

analysis of the neoglycolipid (Fig. 3), in conjunction with methylation analysis of the purified oligosaccharide, this component has been identified as a novel class of ligand (sulfate-containing) for E-selectin. The component is a tetrasaccharide: an equimolar mixture of Le^x- and Le^a-type, sulfated at position 3 of the outer galactose residue:

SO_3–*3*Gal*1–4*/*3*GlcNAc*1–3*Gal
| 1,3/4
Fuc

The binding activity is substantially greater than those of lipid-linked Le^x and Le^a sequences, and it is at least equal to that of the 3'-sialyl-Le^x analogue. Molecular models of the 3'-sialyl-Le^x trisaccharide and the 3'-sulfated analogue indicate similar topologies for the acidic groups in the two oligosaccharides. Thus the sulfate group can substitute for the carboxyl group of the sialic acid.

These results show the power of the neoglycolipid technology to detect hitherto unsuspected ligands among highly heterogeneous oligosaccharides derived from glycoproteins. This is only a beginning, and much further work on this and other glycoproteins is required to establish the range of oligosaccharides recognized by this important adhesion molecule. Little is known about the distribution of the sulfated Le^x and Le^a structures in the body. There has been a report of the occurrence of

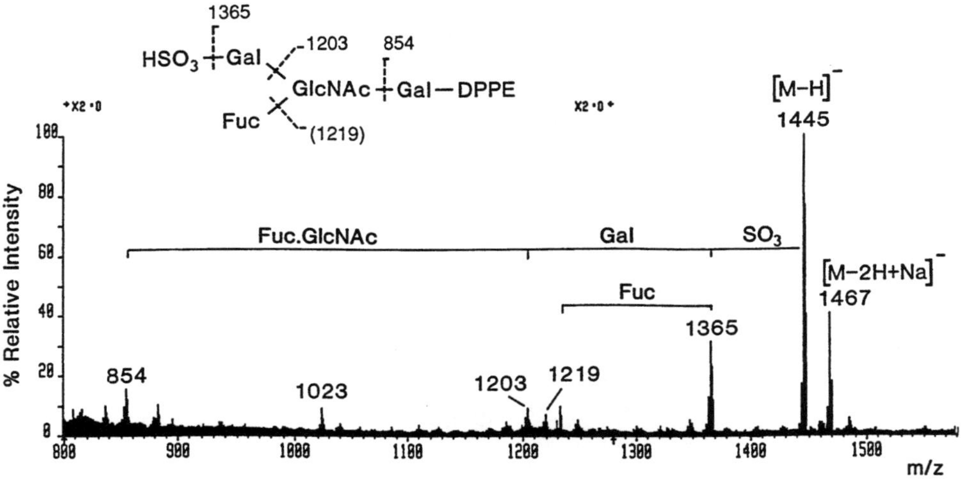

Fig. 3. The negative ion spectrum of the sulfated Le^a and Le^x-type tetrasaccharide. (Yuen et al. 1992)

sulfated Le^x in glycoproteins of the fetal gastrointestinal tract, meconium (Capon et al. 1989). It will be important to raise monoclonal antibodies to each of the two isomeric oligosaccharides and to study their distribution in normal and neoplastic tissues.

Major resources are being directed by several research groups and biotechnology companies towards the design of compounds that are related to saccharide ligands of E-selectin. These would serve as inhibitors of binding, and be useful as therapeutic substances in the management of undesirable or life-threatening instances of E-selectin-mediated adhesion. Initially, the emphasis was on the synthesis of 3'-sialyl-Le^x (Hodgson 1991; Edgington 1992). Our results now render the 3-sulfated analogues strong contenders as inhibitory compounds for E-selectin-mediated adhesion. However, the choice of inhibitors for therapeutic purposes will need to be considered in the context of a database (still to be established) of the ligands for the various carbohydrate-binding proteins in the body. Information so far available on adhesive specifities of L-selectin (see Sect. 4.3) and of the endogenous ligands conglutinin (Mizuochi et al. 1989), mannan-binding protein (Childs et al. 1989) and amyloid P protein (Loveless et al. 1990) indicate that there are some overlaps of binding specificties.

4.3 L-Selectin Binding Studies

We have studied the adhesive specificity of the rat L-selectin molecule whose carbohydrate recognition domain has a greater than 80% homology with that of the human protein (Watanabe et al. 1992) and has been expressed in soluble dimeric form (an L-selectin-IgG chimera) in a baculovirus system. In view of the previously documented occurrence of sulfate on the glycoproteins bound by the L-selectin molecule, it was of special interest to investigate L-selectin binding to the sulfated Le[x]/Le[a] tetrasaccharides described above, in comparison with the sialyl and nonsialyl fuco-oligosaccharides and related analogues.

We observed that the sulfated Le[x]/Le[a] tetrasaccharides are more strongly bound by the L-selectin molecule than 3'-sialyl analogues (Green et al. 1992). There are two further differences from the E-selectin specificity: (1) the nonsulfated fucosylated oligosaccharides LNFP-II and LNFP-III are not bound, and (2) the 3'-sulfated backbone lacking fucose is moderately strongly bound by the L-selectin molecule (Table 2). L-selectin has a special affinity for sulfated oligosaccharides, for we have observed that in the chromatogram overlay assay system it binds to several sulfated glycolipids of the ganglio series derived from brain and from renal tissues (Table 3). There is strong binding to the glycolipids that contain sulfate 3'-linked to terminal galactose or N-acetylgalactosamine or to subterminal galactose, but there is no binding if the sulfated galactose is internally located. L-selectin also binds to the HNK-1 glycolipid which contains a terminal 3'sulfated glucuronic acid. The latter is a well-known component of myelinated nervous tissue and of natural killer cells.

There may be a relationship between the binding to the sulfated acidic oligosaccharides and the recently reported L-selectin binding to tissue sections of the myelineated areas of the central nervous system (Huang et al. 1991), and the distal tubules and capillaries of the kidney medulla (Tamatani et al. 1993). Thus, it will be important to determine whether the sulfated saccharides are accessible for L-selectin-mediated adhesion of infiltrating leucocytes in the intact tissues or whether they become accessible upon tissue damage, and to establish whether L-selectin-mediated leucocyte binding is a mechanism for initiating or perpetuating the pathology of demyelineating diseases, such as the encephalitides and multiple sclerosis, and the nephritides.

The differences observed between L-selectin and E-selectin binding may underlie previous observations that L-selectin binds with a greater intensity than E-selectin to

Table 2. Comparison of E- and L-selectin binding to neoglycolipids

	E[a]	L[b]
NeuAcα2–3Galβ1–3GlcNAcβ1–3Galβ1–4Glc | 1,4 Fucα	++	+
Galβ1–3GlcNAcβ1–3Galβ1–4Glc | 1,4 Fucα	++	–
NeuAcα2–3Galβ1–3GlcNAcβ1–3Galβ1–4Glc	–	–
HSO_3–3Gal1–3/4GlcNAc1–3Gal | 1,4/3 Fuc	++	++
HSO_3–3Gal1–3/4GlcNAc1–3Gal	–	+

[a] E Assayed on transfected CHO cells.
[b] L Assayed as soluble IgGFc chimera.

Table 3. Binding of L-selectin (IgGFc chimera) to sulfated glycolipids on chromatograms

Glycolipids	Sequence	Binding
Sulfatide	(HSO$_3$-3)Gal–Cer	+
SM3	(HSO$_3$-3)Galβ1–4Glc–Cer	+
SM2	GalNAcβ1–4(HSO$_3$-3)Galβ1–4Glc–Cer	+
SM1a	Galβ1–3GalNAcβ1–4(HSO$_3$-3)Galβ1–4Glc–Cer	–
SB2	(HSO$_3$-3)GalNAcβ1–4(HSO$_3$-3)Galβ1–4Glc–Cer	+
SB1a	(HSO$_3$-3)Galβ1–3GalNAcβ1–4(HSO$_3$-3)Galβ1–4Glc–Cer	+
HNK–1	(HSO$_3$-3)GlcUAβ1–3Galβ1–4GlcNAcβ1–3Galβ1–4Glc–Cer	+

a peripheral lymph node vascular glycoprotein designated PNAd, and that E-selectin but not L-selectin binds to the cutaneous lymphocyte antigen CLA (Berg et al. 1992). Thus, predictions may be made on the way binding specificities arise with these two rather "promiscuous" adhesive proteins: (1) through differing affinities of binding to cross-reacting oligosaccharide ligands, and (2) through focal concentrations (clustering) of preferred ligands at "desired" sites on the microvasculature. Direct comparison with the human E-selectin system cannot yet be made, since we have not yet investigated the human L-selectin system. However, from the present findings, prospects seem good for the design of therapeutic compounds that differentially inhibit the binding of these two adhesive molecules.

5 Summary and Conclusions

The concept that carbohydrate differentiation antigens identified using monoclonal antibodies are "area codes" decoded by endogenous lectins that determine cell migration pathways has received support from recent developments in vascular biology. The Le^x and sialyl-Le^x antigens, which are differentiation antigens of human granulocytes and monocytes, are now established ligands for the inducible endothelial cell adhesion molecule E-selectin, which has a crucial role at the initial stages of leucocyte recruitment in inflammation. Similar principles almost certainly apply in blood-borne spread of tumor cells, some of which express at very high levels the above antigens and also the structurally related antigens Le^a, Le^b, and sialyl-Le^a. The blood group H monosaccharide, when present on the Le^a sequence, does not inhibit E-selectin binding, for the resulting Le^b antigen structure supports binding. However, the blood group A and B monosaccharides, when present, as in the ALe^b and BLe^b structures, do not support E-selectin adhesion. Thus the blood group elements may be important variables that influence metastatic potential of human tumor cells.

Since the selectin adhesion system presents opportunities for novel drug designs to control or inhibit undesirable or inappropriate inflammatory reactions, and the metastasis of tumor cells, detailed knowledge of the combining specificities of the adhesion molecules is highly desirable. The neoglycolipid technology offers unique opportunities for such studies since it is applicable to desired structurally characterized or chemically synthesized oligosaccharides as well as to mixtures of unknown oligosaccharides released from glycoproteins. By this technology, a novel class of E-selectin ligands, sulfated Le^a/Le^x, has been identified among oligosaccharides on an ovarian cystadenoma glycoprotein. Binding to these structures is substantially greater than to the Le^x and Le^a sequences, and it is at least equal to that of the sialyl-Le^x analogue. These sulfated oligosaccharides are most potent ligands also for L-selectin. Knowing that there are some cross-reactions among the oligosaccharide ligands for several of the known endogenous carbohydrate-binding proteins, there is need for a database of carbohydrate-binding specificities. This is predicted to be an important requirement for the design of adhesion inhibitors that are "mono-specific" (not cross-reactive) and nontoxic when administered as drugs. For a more up-to-date review of ligands and counter-receptors for the selectins and other endogenous lectins, see Feizi (1993).

Acknowledgments. Supported by the Medical Research Council and grants from the Leukaemia Research Fund, The Arthritis and Rheumatism Council, and Genetics Institute Inc.

References

Brandley BK, Swiedler SJ, Robbins PW (1990) Carbohydrate ligands of the LEC cell adhesion molecules. Cell 63:861–863

Berg EL, Robinson MK, Mansson O, Butcher EC, Magnani JL (1991) A carbohydrate domain common to both sialyl Le^a and sialyl-Le^x is recognized by the endothelial cell leukocyte adhesion molecule ELAM-1. J Biol Chem 226:14869–14872

Berg EL, Magnani J, Warnock RA, Robinson MK, Butcher EC (1992) Comparison of L-selectin and E-selectin ligand specificities: the L-selectin can bind the E-selectin ligands sialyl-Le^x and sialyl Le^a. Biochem Biophys Res Commun 184:1048–1055

Capon C, Leroy Y, Wheruszeski J-M, Ricardi G, Strecker G, Montreuil J, Fournet B (1989) Structures of 0-glycosidically linked oligosaccharides isolated from human meconium glycoproteins. Eur J Biochem 182:139–152

Childs RA, Kapadia A, Feizi T (1980) Expression of blood group I and i active carbohydrate sequences on cultured human and animal cell lines assessed by radioimmunoassay with monoclonal cold agglutinins. Eur J Immunol 10:379–384

Childs RA, Drickamer K, Kawasaki T, Thiel S, Mizuochi T, Feizi T (1989) Neoglycolipids as probes of oligosaccharide recognition by recombinant and natural mannose-binding proteins of the rat and man. Biochem J 262:131–138

Childs RA, Wright JR, Ross GF, Yuen C-T, Lawson AM, Chai W, Drickamer K, Feizi T (1992) Specificity of lung surfactant protein SP-A for both the carbohydrate and the lipid moieties of certain neutral glycolipids. J Biol Chem 267:9972–9979

Edgington SM (1992) How sweet it is: selectin-mediating drugs. Bio/Technology 10:383–389

Feizi T (1981) Carbohydrate differentiation antigens. Trends Biochem Sci 6:333–335

Feizi T (1983) In: Evered D, Whelan J (eds) Fetal antigens and cancer. Ciba Foundation Symposium (1982) Vol 96 pp 216–221

Feizi T (1984) Monoclonal antibodies reveal saccharide structures of glycoproteins and glycolipids as differentiation and tumour-associated antigens. Biochem Soc Trans 12:545–549

Feizi T (1985) Demonstration by monoclonal antibodies that carbohydrate structures of glycoproteins and glycolipids are onco-developmental antigens. Nature 314:53–57

Feizi T (1989 Glycoprotein oligosaccharides as recognition structures. In: Bock G, Harnett S (eds) Carbohydrate recognition in cellular function. Ciba Foundation Symposium 1988, Vol 145, Wiley, Chichester, pp 62–79

Feizi T (1991a) Carbohydrate differentiation antigens: probable ligands for cell adhesion molecules. Trends Biochem Sci 16:84–86

Feizi T (1991b) Cell-cell adhesion and membrane glycosylation. Curr Opin Struct Biol 1:766–770

Feizi T (1993) Oligosaccharides that mediate cell-cell adhesion. Curr Opin Struct Biol 3:701–710

Feizi T, Childs RA (1987) Carbohydrates as antigenic determinants of glycoproteins. Biochem J 245:1–11

Feizi T, Stoll MS, Yuen C-T, Chai W, Lawson AM (1993) Neoglycolipids – probes of oligosaccharide structure, antigenicity and function. Methods Enzymol 230:484–519

Foxall C, Watson SR, Dowbenko D, Fennie C, Lasky LA, Kiso M, Hasegawa A, Asa D, Brandley BK (1992) The three members of the selectin receptor family recognize a common carbohydrate epitope, the sialyl Lewis^{x} oligosaccharide. J Cell Biol 117:895–902

Gooi HC, Feizi T, Kapadia A, Knowles BB, Solter D, Evans MJ (1981) Stage-specific embryonic antigen SSEA-1 involves α1–3 fucosylated type 2 blood group chains. Nature 292:156–158

Gooi HC, Thorpe SJ, Hounsell EF, Rumpold H, Kraft D, Förster O, Feizi T (1983) Marker of peripheral blood granulocytes and monocytes of man recognized by two monoclonal antibodies VEP8 and VEP 9 involves the trisaccharide 3-fucosyl-N-acetyllactosamine. Eur J Immunol 13:306–312

Green PJ, Tamatani T, Watanabe T, Miyasaka M, Hasegawa A, Kiso M, Yuen C-T, Stoll MS, Feizi T (1992) High affinity binding of the leucocyte adhesion molecule L-selectin to 3'-sulfated-Le^{a} and -Le^{x} oligosaccharides and the predominance of sulphate in this interaction demonstrated by binding studies with a series of lipid-linked oligosaccharides. Biochem Biophys Res Commun 188:244–251

Harlan JM, Liu DY (eds) Adhesion: its role in inflammatory disease. WH Freeman, New York

Hodgson J (1992) Carbohydrate-based therapeutics. Bio/Technology 9:609–613

Huang K, Geoffroy JS, Singer MS, Rosen SD (1991) A lymphocyte homing receptor (L-selectin) mediates the in vitro attachment of lymphocytes to myelinated tracts of the central nervous system. J Clin Invest 88:1778–1783

Imai Y, Singer MS, Fennie C, Lasky LA, Rosen SD (1991) Identification of a carbohydrate-based endothelial ligand for a lymphocyte homing receptor. J Cell Biol 113:1213–1221

Imai Y, Lasky LA, Rosen SD (1993) Sulphation requirement for GlyCAM-1, an endothelial ligand for L-selectin. Nature 361:555–557

Köhler G, Milstein C (1975) Continuous cultures of fused cells secreting antibody of predefined specificity. Nature (Lond) 256:495–497

Larkin M, Ahern TJ, Stoll MS, Shaffer M, Sako D, O'Brian J, Yuen C-T, Lawson AM, Childs RA, Barone KM, Langer-Safer PR, Hasegawa A, Kiso M, Larsen GR, Feizi T (1992)

Spectrum of sialylated and nonsialylated fuco-oligosaccharides bound by the endothelial leukocyte adhesion molecule E-selectin. Dependence of the carbohydrate-binding activity on E-selectin density. J Biol Chem 267:13661–13668

Lasky LA, Singer MS, Dowbenko D, Imai Y, Henzel WJ, Grimley C, Fennie C, Gillett N, Watson SR, Rosen SD (1992) An endothelial ligand for L-selectin is a novel mucin-like molecule. Cell 69:927–938

Lawson AM, Chai W, Cashmore GC, Stoll MS, Hounsell EF, Feizi T (1990) High-sensitivity structural analyses of oligosaccharide probes (neoglycolipids) by liquid-secondary-ion mass spectrometry. Carbohydr Res 200:47–57

Loveless RW, Bellairs R, Thorpe SJ, Page M, Feizi T (1990) Developmental patterning of the carbohydrate antigen FC10.2 during early embryogenesis in the chick. Development 108:97–106

Loveless RW, Floyd-O'Sullivan G, Raynes JG, Yuen C-T, Feizi T (1992) Human serum amyloid P is a multispecific adhesive protein whose ligands include 6-phosphorylated mannose and the 3-sulphated saccharides galactose, N-acetylglucosamine and glucuronic acid. EMBO J 11:813–819

Lowe JB, Stoolman LM, Nair RP, Larsen RD, Berhend TL, Marks RM (1990) ELAM-1-dependent cell adhesion to vascular endothelium determined by a transfected human fucosyltransferase cDNA. Cell 63:475–484

Macher BA, Buehler J, Scudder P, Knapp W, Feizi T (1988) A novel carbohydrate differentiation antigen on fucogangliosides of human myeloid cells recognized by monoclonal antibody VIM-2. J Biol Chem 263:10186–10191

Magnani JL (1984) Carbohydrate differentiation and cancer-associated antigens detected by monoclonal antibodies. Biochem Soc Trans 12:543–545

Mizuochi T, Loveless RW, Lawson AM, Chai W, Lachmann PJ, Childs RA, Thiel S, Feizi T (1989) A library of oligosaccharide probes (neoglycolipids) from N-glycosylated proteins reveals that conglutinin binds to certain complex type as well as high-mannose typ oligosaccharide chains. J Biol Chem 264:13834–13839

Moore KL, Stults NL, Siaz S, Smith DF, Cummings RD, Varki A, McEver RP (1992) Identification of a specific glycoprotein ligand for P-selectin (CD62) on myeloid cells. J Cell Biol 118:445–456

Solter D, Knowles BB (1978) Monoclonal antibody defining a stage-specific mouse embryonic antigen (SSEA-1). Proc Natl Acad Sci USA 75:5565–5569

Stoll MS, Mizuochi T, Childs RA, Feizi T (1988) Improved procedure for the construction of neoglycolipids having antigenic and lectin-binding activities from reducing oligosaccharides. Biochem J 256:661–664

Stoll MS, Hounsell EF, Lawson AM, Chai W, Feizi T (1990) Microscale sequencing of 0-linked oligosaccharides using mild periodate oxidation of alditols, coupling to phospholipid and TLC-MS analysis of the resulting neoglycolipids. Eur J Biochem 189:499–507

Takada A, Ohmori K, Takahashi N, Tsuyoka K, Yago A, Zenita K, Hasegawa A, Kannagi R (1991) Adhesion of human cancer cells to vascular endothelium mediated by a carbohydrate antigen, sialyl Lewis A. Biochem Biophys Res Commun 179:713–719

Tamatani T, Kuida K, Watanabe T, Koike S, Miyasaka M (1993) Molecular mechanisms underlying lymphocyte recirculation: III. Characterization of the LECAM-1 (L-selectin)-dependent adhesion pathway in rats. J Immunol 150:1735–1745

Tang PW, Gooi HC, Hardy M, Lee YC, Feizi T (1985) Novel approach to the study of the antigenicities and receptor functions of carbohydrate chains of glycoproteins. Biochem Biophys Res Commun 132:474–480

Thorpe SJ, Feizi R (1984) Species differences in the expression of carbohydrate differentiation antigens on mammalian blood cells revealed by immunofluorescence with monoclonal antibodies. Biosci Reps 4:673–685

Uemura K-I, Macher BA, DeGregorio M, Scudder P, Buehler J, Knapp W, Feizi T (1985) Glycosphingolipid carriers of carbohydrate antigens of human myeloid cells recognized by monoclonal antibodies. Biochem Biophys Acta 846:26–36

Watanabe T, Song Y, Hirayama Y, Tamatani T, Kuida K, Miyasaka M (1992) Sequencing and expression of a rat cDNA for LECAM-1. Biochem Biophys Acta 1131:321–324
Yuen C-T, Lawson AM, Chai W, Larkin M, Stoll MS, Stuart AC, Sullivan FX, Ahern TJ, Feizi T (1992) Novel sulfated ligands for the cell adhesion molecule E-selectin revealed by the neoglycolipid technology among 0-linked oligosaccharides on an ovarian cystadenoma glycoprotein. Biochemistry 31:9126–9131

The Biologic Significance of Glycoprotein Hormone Oligosaccharides

J. U. Baenziger[1]

1 Introduction

The glycoprotein hormones lutropin (LH), follitropin (FSH), thyrotropin (TSH), and chorionic gonadotropin (CG) are dimeric proteins. They consist of a common α subunit and highly homologous, but hormone-specific, β subunits (Fig. 1). Even though the structures of the glycoprotein hormones are closely related at the primary, secondary, and tertiary levels, the structures of their Asn-linked oligosaccharides differ. LH and TSH bear oligosaccharides which terminate with the unique sequence SO_4-4GalNAcβ1,4GlcNAcβ1,2Manα (S4GGnM) whereas FSH and CG bear oligosaccharides which terminate with the sequence sialic acid α2,3/6Galβ1,4GlcNAcβ1,2Manα (Baenziger and Green 1991; Green and Baenziger 1988a, b; Green et al. 1985b; Endo et al. 1979; Mizuochi and Kobata 1980). The synthesis of these sulfated oligosaccharides reflects the activity of two highly specific enzymes, a GalNAc-transferase and a sulfotransferase. The GalNAc-transferase recognizes features encoded within

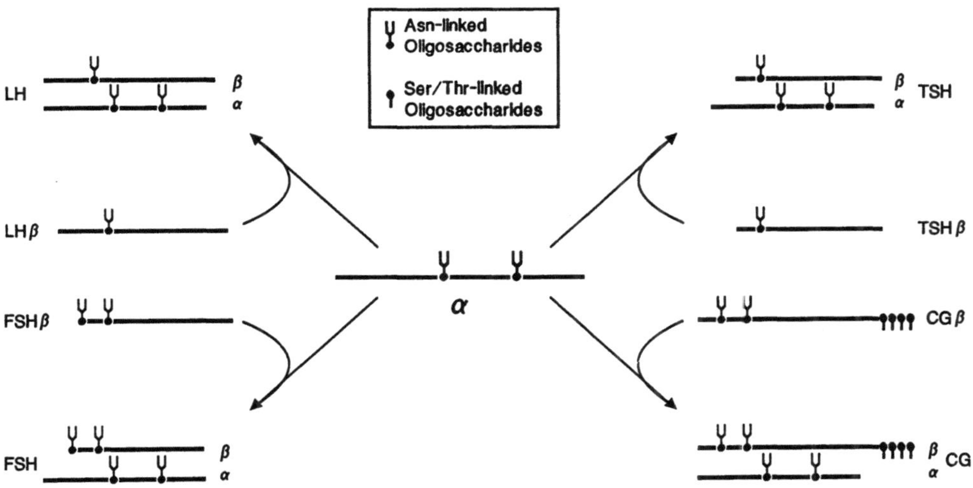

Fig. 1. The glycoprotein hormon family. α and β subunits associate to form distinct hormone heterodimers. The common α subunit is capable of combining with any of four hormone-specific β subunits to form unique α-β heterodimers

[1] Washington University School of Medicine, Department of Pathology, St. Louis, MO 63110, USA.

44. Colloquium Mosbach 1993
Glyco- and Cellbiology

the peptide of the hormone in addition to the oligosaccharide acceptor, whereas the sulfotransferase recognizes only the oligosaccharide acceptor. The sulfated oligosaccharides on LH are recognized by a receptor found in liver, which mediates the rapid clearance of LH from the blood. The expression of the GalNAc- and sulfotransferases is hormonally regulated and the sulfated oligosaccharides produced play a key role in the expression of LH bioactivity in vivo by controlling its circulatory half-life following release into the blood.

2 Synthesis of Oligosaccharides Terminating with SO_4-4GalNAcβ1,4GlcNAcβ1,2Manα.

The synthetic pathway for the sulfated and sialylated oligosaccharides found on the glycoprotein hormones is illustrated in Fig. 2. The structures enclosed within the box are intermediates in the synthetic pathway which are common for sulfated and sialylated oligosaccharides. Since identical oligosaccharide intermediates act as acceptors for the addition of GalNAc and Gal by their respective glycosyltransferase, one of the two glycosyltransferases must display specificity for the underlying peptide.

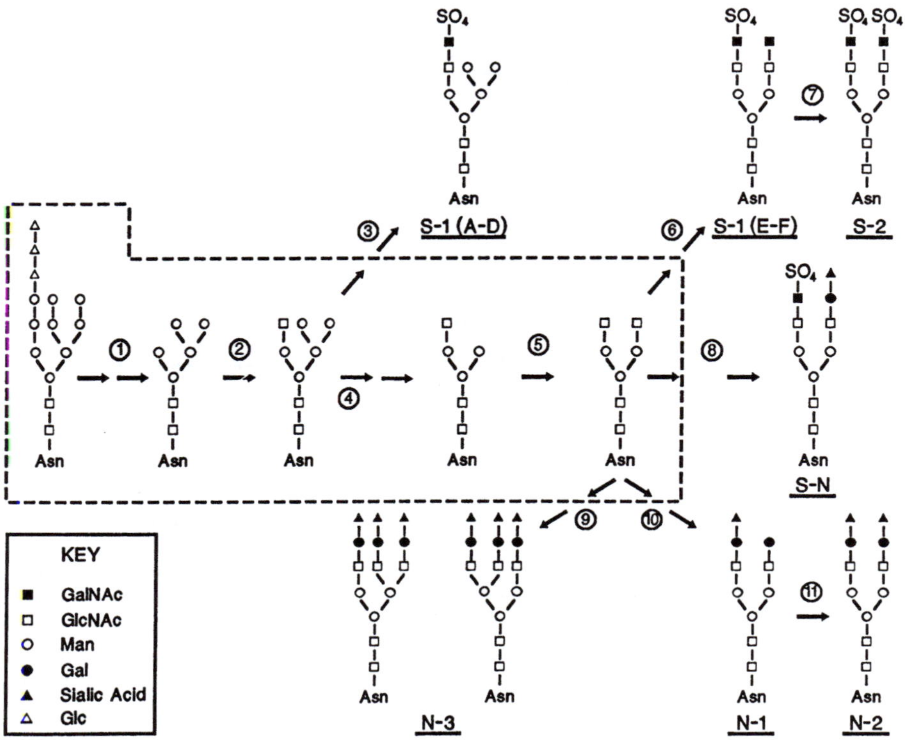

Fig. 2. Proposed pathway for the synthesis of sulfated and sialylated oligosaccharides on the pituitary glycoprotein hormones

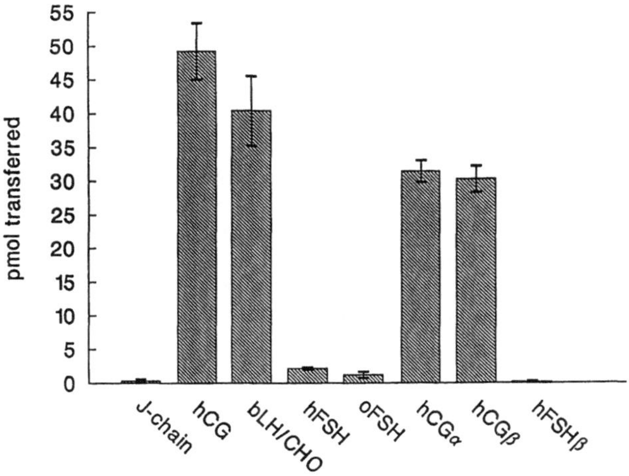

Fig. 3. Comparison of glycoproteins as substrates for the GalNAc-transferase. Each glycoprotein was digested with neuraminidase and β-galactosidase to produce oligosaccharides with the structure of the product of reaction 5 in Fig. 2. Assays contained 4 μM substrate, 720 μM UDP-GalNAc, and 385 μU/ml of GalNAc-transferase activity. *J-chain* contains one Asn-linked oligosaccharide, whereas *hCG*, *hFSH* and *oFSH* contain four, and *bLH/CHO* contains three. The isolated subunits *hCGα*, *hCGβ*, and *hFSHβ* each contain two Asn-linked oligosaccharides

This was established by comparing the glycoproteins in Fig. 3, each bearing the identical oligosaccharide acceptor $GlcNAc_2Man_3GlcNAc_2Asn$, as acceptors for Gal or GalNAc addition by transferase present in bovine pituitary membranes. Galactose was added to each of these glycoprotein acceptors with equal efficiency, confirming that galactosyltransferase does not display protein specificity. In contrast, GalNAc was transferred to the oligosaccharides acceptors on hCG, LH, hCGα, and hCGβ with much greater efficiency than to the identical acceptors on J-chain, oFSH, hFSH, and hFSHβ (Fig. 3) (Smith and Baenziger 1990), Thus, a GalNAc-transferase is present in pituitary membranes which discriminates between identical oligosaccharide acceptors on LH, CG, and other glycoproteins including FSH.

We have examined the specificity of this GalNAc-transferase (Smith and Baenziger 1988, 1990, 1992) and determined that it recognizes a tripeptide motif found in the α subunit as well as the β subunits of LH and hCG. This tripeptide motif consits of a ProXaaArg/Lys (PXR/K) located 6–9 residues N-terminal to a glycosylated Asn (Fig. 4) and is absent from the β subunit of FSH due to an amino terminal deletion within the FSHβ gene as compared to LHb and CGβ. Furthermore, combination of FSHβ with a subunit prevents recognition of the α subunit tripeptide motif by the GalNAc-transferase (Fig. 3). The apparent K_m for GalNAc addition to the synthetic intermediate $GlcNAc_2Man_3GlcNAc_2Asn$ on proteins containing the tripeptide motif is 4–13 μM, as compared to 1–2 mM for addition to the same oligosaccharide acceptor on proteins which do not contain the tripeptide motif (Smith and Baenziger 1990).

Thus, the tripeptide motif serves to lower the apparent K_m for GalNAc addition to the oligosaccharide acceptor but is not essential for transfer. In contrast to the Gal-

hCGβ	S-K-E-P-L-R-P-R-C-R-P-I-N-A-T-L-A-V-E-K
bLHβ	S-R-G-P-L-R-P-L-C-Q-P-I-N-A-T-L-A-A-E-K
hFSHβ	N-S-C-E-L-T-N-I-T-I-A-I-E-K
hCGα	R-A-Y-P-T-P L R-S-K-K-T-M-L-V-Q-K-N-V-T-S-E

Fig. 4. Alignment of peptides known to be recognized by the glycoprotein hormone-specific GalNAc-transferase and comparison with the amino terminal sequence of hFSHβ. The amino acid sequences represented in the single letter code are aligned with respect to glycosylated Asn residues *(dashed box)*. The ProXaaArg/Lys motif required for recognition by the GalNAc-transferase is *underlined*. The amino acid sequences correspond to amino acids 1–20 for *hCG*β, 1–20 for *bLH*β, 1–14 for *hFSH*β, and 35–56 for human α subunit *(hCG*α*)*

NAc-transferase, the sulfotransferase requires only the presence of the trisaccharide sequence GalNAcβ1,4GlcNAcβ1,2Manα (GGnM) for transfer of sulfate to the 4-hydroxyl of the GalNAc (Skelton et al. 1991; Green et al. 1985a). The PXR/K-specific GalNAc-transferase and the GGnM-4-sulfotransferase are both expressed in pituitary, whereas neither is detected in human placental tissue, accounting for the absence of sulfated oligosaccharides on hCG despite the presence of the tripeptide recognition motif (Green et al. 1985a; Smith and Baenziger 1988).

3 Levels of Galnac- and Sulfo-Transferase Are Hormonally Regulated in the Pituitary

Glycoprotein hormones such as LH are highly heterogeneous with multiple isoforms which differ in isoelectric point (Keel 1989). These isoforms have been attributed to differences in the terminal glycosylation. Differences in the potency of these isoforms for LH/CG-receptor activation have led to the proposal that terminal glycosylation of LH oligosaccharides modulates is potency. Since the isoforms present in serum have been observed to change with hormonal state, there may be a relationship between hormonal state and levels of glycosyltransferase expression. Differences in the efficiency of GalNAc or Gal addition to LH oligosaccharides would have the effect of altering the isoforms of LH by changing the porportions of terminal glycosylation with SO_4 and sialic acid. Levels of LH synthesis and secretion by gonadotrophs during the ovulatory cycle change dramatically, particularly at the time of the pre-ovulatory LH "surge". These marked changes in levels of LH synthesis could result in a shift in the proportion of oligosaccharides on LH, terminating with GalNAc-4-SO_4 as opposed to sialic acid-Gal.

Using sensitive and highly specific assays for the PXR/K-specific GalNAc-transferase (Mengeling et al. 1991) and GGnM-4-sulfotransferase (Smith and Baenziger 1988), we examined the effect of ovariectomy and ovariectomy with estrogen replacement on the levels of transferase expression in pituitary. LH levels in the pituitary and serum rise three- to fivefold following ovariectomy. Similary, the levels of PXR/K-specific GalNAc-transferase and GGnM-4-sulfotransferase in rat pituitaries increase in response to ovariectomy. Administration of exogenous 17β-estradiol to ovariectomized rats results in a return of LH to basal levels in pituitary. Likewise, the levels of PXR/K-specific GalNAc-transferase and GGnM-4-sulfotransferase return to basal levels in response to exogenous estradiol. The PXR/K-specific GalNAc-transferase and GGnM-4-sulfotransferase respond to estrogen levels in the same manner and to a similiar extent as LH. Since the levels of the PXR/K-specific GalNAc-transferase and the GGN-4-sulfotransferase in gonadotrophs are regulated by estrogen in a similar manner as their substrate, LH, this suggests that the Asn-linked oligosaccharides on LH will terminate with GalNAc-4-SO_4 regardless of the levels of LH synthesis. In support of this conclusion, we have found that the proportion of sulfated oligosaccharides on LH is not altered following ovariectomy (Dharmesh and Baenziger 1993).

4 The Functional Significance of Sulfated Oligosaccharides on LH

The presence of unique sulfated oligosaccharides on LH from a number of different animal species raises the question of its functional significance. The importance of this question is amplified by the knowledge that expression of the GalNAc- and sulfotransferases responsible for the synthesis of these sulfated oligosaccharides is modulated by estrogen levels, assuring that LH oligosaccharides terminate with the S4GGnM sequence at all levels of LH synthesis. We have examined three potential functions which could be attributed to these unique oligosaccharides: (1) directing the sorting of LH to the appropriate storage granule, (2) modulating the binding to or activation of the LH/GG receptor, or (3) controlling the circulatory half-life of LH following release into the blood.

Terminal glycosylation of LH can be altered in cultured cells by the use of selective inhibitors of oligosaccharide processing and by inhibition of sulfate addition. Using such agents, we have not been able to demonstrate that either the presence of sulfate or the structure of the oligosaccharides on LH play an essential role in directing the hormone to granules in primary cultures of bovine gonadotrophs. Even though LH and FSH are synthesized within the same cells, they are segregated to separate granules within the gonadotroph. Terminal glycosylation does not appear to play an essential role in directing LH into the regulated pathway of secretion; it may, however, have an impact on the segregation of LH and FSH to seperate granules.

We have examined the impact of terminal glycosylation and binding to and activation of the LH/CG receptor using the cultured cell line, MA-10, which expresses the LH/CG-receptor (Smith et al. 1990; Baenziger et al. 1992). Removal of the terminal sulfate from native bovine LH has no impact on binding to the LH/CG receptor, production of cyclic AMP, or production of progesterone. Chinese hamster ovary cells (CHO) express neither the GalNAc- nor the sulfotransferase. As a result, recombinant bovine LH expressed in CHO cells bears oligosaccharides terminating with sialic acid α2,3Gal rather than GalNAc-4-SO_4. Recombinant LH is less potent than native LH; however, following enzymatic removal of the terminal sialic acid, recombinant LH has the same potency as native LH. Thus, the presence of terminal sialic acid reduces the potency of LH whereas the presence of terminal sulfate has no effect at the level of LH/CG receptor activation in vitro. Since both sulfate and sialic acid are negatively charged, the reduced potency seen with sialic acid is more likely to be related to steric effects than to charge. Furthermore, the detailed structural features of the oligosaccharides on LH do not appear to have a marked impact on binding to or activation of the LH/CG receptor.

Terminal glycosylation of the Asn-linked oligosaccharides on LH has a major impact on its circulatory half-life and site of clearance from the blood (Fiete et al. 1991; Baenziger et al. 1992). Native LH is cleared from the blood five fold more rapidly than recombinant LH bearing oligosaccharides terminating with sialic acid-Gal (Fig. 5). Since the only difference between native LH and recombinant LH resides in the identity of their terminal sugars, this indicates that the GalNAc-4-SO_4 is responsible for the rapid clearance of native LH. We subsequently demonstrated that LH is rapidly removed from the circulation by a receptor which is present in hepatic endothelial/Kupffer cells. This receptor recognizes oligosaccharides with the terminal sequence SO_4-4GalNAcβ1,4GlcNAcβ1,2Manα. But not oligosaccharides terminating with the sequence SO_4-3GalNAcβ1,4GlcNAcβ1,2Manα. Fucoidin, but not other sulfated polysaccharides such as heparin, chondroitin-sulfate, and dextran sulfate, inhibit binding to the receptor. Thus, the receptor is highly specific, requiring GalNAc-4-SO_4 for recognition. LH is bound by the S4GGnM-receptor with an apparent K_m of 1.6×10^{-7}M and is rapidly internalized and degraded. Hepatic endothelial cells express > 500 000 S4GGnM-binding sites per cell at their surface, indicating that this receptor system has the capacity to remove from the blood large amounts of LH and other glycoproteins bearing oligosaccharides terminating with S4GGnM.

What is the functional significance of the S4GGnM-receptor and the sulfated oligosaccharides on LH? We have compared the ability of different forms of LH to stimulate ovulation in mice (Baenziger et al. 1992). The potency of LH to stimulate ovulation correlated with both its affinity for the LH/CG receptor and its circulatory half-life. As a result, the level of circulating LH which must be attained to stimulate ovulation is directly related to its rate of clearance from the blood as well as to its affinity for the LH/CG receptor. A characteristic of LH secretion is its pulsatile pattern

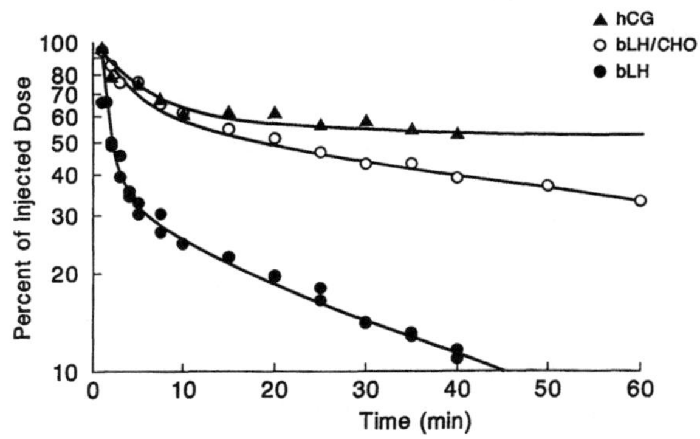

Fig. 5. Plasma clearance rates for bovine LH *(bLH)*, recombinant bovine LH *(bLH/CHO)* and *hCG*. Hormones labeled with ^{125}I (0.5 – 2.0 x 10^6 cpm, 50–200 ng) were injected into heparinized anesthetized rats. Samples (200 µl) were drawn from the carotid artery at the times indicated. Radioactivity is expressed as the percent of the injected dose per ml of plasma

(Veldhuis et al. 1985; Veldhuis and Johnson 1988), which requires rapid clearance of released LH from the blood. Since the LH/CG receptor is downregulated following ligand binding (Wang et al. 1991; Rodriguez et al. 1992), the pulsatile pattern of appearance in the blood may reflect a requirement to maintain the maximum number of LH/CG receptors in the ground state at the time of exposure to a new pulse of LH. Thus, reduced potency due to rapid clearance from the blood would be compensated for by maximal activation of cAMP production at the level of the LH/CG receptor.

Additional evidence for the importance of terminal glycosylation for expression of LH bioactivity in vivo comes from a comparison of equine LH and CG which have identical α and β subunit peptides but are synthesized in pituitary and placenta, respectively. Equine LH bears oligosaccharides terminating with SO_4-4GalNAcβ-1,2Manα whereas equine CG oligosaccharides terminate with sialic acidα-Galβ-1,4GlcNAcβ1,2Manα (Smith et al. 1993). Equine LH, but not equine CG, is recognized by the S4GGnM-receptor and is rapidly cleared from the circulation. Thus, the major difference between equine CG and LH lies in their glycosylation and its affect on circulatory half-life, suggesting that rapid clearance is essential for ovulation but not for maintenance of pregnancy.

Oligosaccharides terminating with SO_4-4GalNAcβ1,5GlcNAcβ1,2Manα are not confined to the glycoprotein hormones. Proopiomelanocortin produced in AtT-20 cells (Skelton et al. 1992) and recombinant tissue factor pathway inhibitor expressed in 293 cells (Smith et al. 1992) both bear oligosaccharides terminating with the S4GGnM sequence. Both proteins contain a PXR/K recognition motif and both cell lines express the GalNAc- and sulfotransferases. In addition, we have detected the PXR/K-specific GalNAc-transferase and the GGnM-4-sulfotransferase in tissues such as salivary gland, kidney, and liver (Dharmesh et al. 1993). It is likely that a number of additional glycoproteins will be found to bear oligosaccharides terminating with S4GGnM. Although the functional significance of these oligosaccharide structures on other glycoproteins may differ from that of LH, it is likely that they will play an equally important biologic role for these glycoproteins.

5 Conclusions

The sulfated oligosaccharides found on LH are critical for the expression of its in vivo bioactivity. The sulfated oligosaccharides are recognized by an abundant receptor in hepatic endothelial/Kupffer cells which mediates the rapid clearance of LH from the circulation. The GalNAc-transferase responsible for synthesis of these structures recognizes a tripeptide motif, ProXaaArg/Lys, which is found on both the α and β subunits of LH. In contrast, the sulfotransferase requires only the trisaccharide sequence GalNAcβ1,4GlcNAcβ1,2Manα for transfer. Both transferases are regulated by estrogen levels in a manner similar to that of LH, assuring that LH bears oligosaccharides terminating with SO_4-4GalNAcβ1,4GlcNAcβ1,2Manα and will be cleared by the S4GGnM-receptor.

References

Baenziger JU, Kumar S, Brodbeck RM, Smith PL, Beranek MC (1992) Circulatory half-life but not interaction with the lutropin/chorionic gonadotropin receptor is modulated by sulfation of bovine lutropin in oligosaccharides. Proc Natl Acad Sci USA 89:334–338

Baenziger JU, Green ED (1991) Structure, synthesis, and function of the asparagine-linked oligosaccharides on pituitary glycoprotein hormones. In: Ginsberg B, Robbins PW (eds) Biolgoy of carbohydrates, vol 3. JAI Press Ltd, London, pp 1–46

Dharmesh S, Skelton TP, Baenziger JU (1993) Co-ordinate and restricted expression of the PXR/K-specific GalNAc-transferase and the GalNAcβ1,4GlcNAcβ1,2Manα-4-sulfotransferase. (unpubl)

Dharmseh SM, Baenziger JU (1993) Co-ordinate regulation of lutropin, GalNAc- and sulfotransferase in pituitary. (unpubl)

Endo Y, Yamashita K, Tachibana Y, Tojo S, Kobata A (1979) Structures of the Asparagine-linked sugar chains of human chrionic gonadotropin. J Biochem (Tokyo) 85:669–679

Fiete D, Srivastava V, Hindsgaul O, Baenziger JU (1991) A hepatic reticuloendothelial cell receptor specific for SO_4-4GalNAcβ1,4GlcNAcβ1,2Manα that mediates rapid clearance of lutropin. Cell 67:1103–1110

Green ED, Morishima C, Boime I, Baenziger JU (1985a) Structural requirements for sulfation of asparagine-linked oligosaccharides of lutropin. Proc Natl Acad Sci USA 82:7850–7854

Green ED, Van Halbeek H, Boime I, Baenziger JU (1985b) Structural elucidation of the disulfated oligosaccharide from bovine lutropin. J Biol Chem 260:15623–15630

Green ED, Baenziger JU (1988a) Asparagine-linked oligosaccharides on lutropin, follitropin, and thryotropin: I. Structural elucidation of the sulfated and sialylated oligosaccharides on bovine, ovine, and human pituitary glycoprotein hormones. J Biol Chem 263:25–35

Green ED, Baenziger JU (1988b) Asparagine-linked oligosaccharides on lutropin, follitropin, and thyrotropin: II. Distributions of sulfated and sialylated oligosaccharides on bovine, ovine, and human glycoprotein hormones. J Biol Chem 263:36–44

Keel BA (1989) In: Keel BA, Grotjan HE (eds) Microheterogeneity of glycoprotein hormones. CRC Press Boca Raton, pp 203–215

Mengeling BJ, Smith PL, Stults NL, Smith DF, Baenziger JU (1991) A microplate assay for analysis of solution phase glycosyltransferase reactions: determination of kinetic constants. Anal Biochem 199:286–292

Mizuochi T, Kobata A (1980) Different asparagine-linked sugar chains on the two polypeptide chains of human chorionic gonadotropin. Biochem Biophys Res Commun 97:772–778

Rodriguez MC, Xie Y-B, Wang H, Collison K, Segaloff DL (1992) Effects of truncations of the cytoplasmic tail of the luteinizing hormone/chorionic gonadotropin receptor on receptor-mediated hormone internalization. Mol Endocrinol 6:327–336

Skelton TP, Hooper LV, Srivastava V, Hindsgaul O, Baenziger JU (1991) Characterization of a sulfotransferase responsible for the 4-O-sulfation of terminal β-N-acetyl-D-galactosamine on asparagine-linked oligosaccharides of glycoprotein hormones. J Biol Chem 266:17142–17150

Skelton TP, Kumar S, Smith PL, Beranek MC, Baenziger JU (1992) Proopiomelanocortin synthesized by corticotrophs bears asparagine-linked oligosaccharides terminating with SO_4-4GalNAcβ1,4GlcNAcβ1,2Manα. J Biol Chem 267:12998–13006

Smith PL, Kaetzel D, Nilson J, Baenziger JU (1990) The sialylated oligosaccharides of recombinant bovine lutropin modulate hormone bioactivity. J Biol Chem 265:874–881

Smith PL, Skelton TP, Fiete D, Dharmesh SM, Beranek MC, MacPhail L, Broze GJ, Jr., Baenziger JU (1992) The asparagine-linked oligosaccharides on tissue factor pathway inhibitor terminate with SO_4-4GalNAcβ1,4GlcNAcβ1,2Manα. J Biol Chem 267:19140–19146

Smith PL, Bousfield GS, Kumar S, Fiete D, Baenziger JU (1993) Equine lutropin and chorionic gonadotropin bear oligosaccharides terminating with SO_4-4-GalNAc and sialic acid α2,3Gal respectively. J Biol Chem 268:795–802

Smith PL, Baenziger JU (1988) A pituitary N-acetylgalactosamine transferase that specifically recognizes glycoprotein hormones. Science 242:930–933

Smith PL, Baenziger JU (1990) Recognition by the glycoprotein hormone-specific N-acetylgalactosaminetransferase is independent of hormone native conformation. Proc Natl Acad Sci USA 87:7275–7279

Smith PL, Baenziger JU (1992) Molecular basis of recognition by the glycoprotein hormon-specific N-acetylgalactosamine-transferase. Proc Natl Acad Sci USA 89:329–333

Veldhuis JD, Evans WS, Johnson ML, Wills MR, Rogol AD (1986) Physiological properties of the luteinizing hormone pulse signal: impact of intensive and extended venous sampling paradigms on its characterization in healthy men and women. J Clin Endocrinol Metab 62:881–891

Veldhuis JD, Carlson ML, Johnson ML (1987) The pituitary gland secretes in bursts: appraising the nature of glandular secretory impulses by simultaneous multiple-parameter deconvolution of plasma hormone concentrations. Proc Natl Acad Sci USA 84:7686–7690

Veldhuis JD, Johnson ML (1988) A novel general biophysical model for simulating episodic endocrine gland signaling. Am J Physiol 255:E749–E759

Wang H, Segaloff DL, Ascoli M (1991) Lutropin/choriogonadotropin downregulates its receptor by both receptor-mediated endocytosis and a cAMP-dependent reduction in receptor mRNA. J Biol Chem 266:780–785

Biosynthesis of Antibodies and Molecular Chaperones

A. Cremer, M. R. Knittler, and I. G. Haas[1]

1 Introduction

1.1 A Brief Historical Excursion

The idea that antibody molecules may require helper molecules to gain their functional conformation is not new, but originated at a time when only little was known about the structure of antibodies. Landsteiner and others had shown that large numbers of antibodies could be produced by the vertebrate host although it was not clear how these molecules could react with so many different antigens (Landsteiner 1945). As an expansion of the original instruction theories, which said that antibodies were directly molded by the antigen, Burnet proposed in 1941 that it was the function of the antigen to stimulate an adaptive modification of those enzymes necessary for antibody synthesis, such that a unique protein molecule with a specific binding site would result (Burnet 1941). However, these theories were proven false by the demonstration that denatured antibodies, which were allowed to refold in vitro, did not require the presence of antigen to obtain their specific binding property (Haber 1964; Whitney and Tanford 1965). From Anson's observations that denatured hemoglobin could be converted back into native protein (Anson 1945), it became obvious that the amino acid sequence contained all the information necessary to acquire the correct three-dimensional structure of the corresponding protein. Nevertheless, it is not yet possible to predict a three-dimensional structure directly from primary sequence information (Jaenicke 1988). The notion, however, that protein folding is a spontaneous self-organized process was so strong that – at least in the beginning of the chaperone research era – many molecular biologists refused to believe in the existence of special molecules that support such processes in the cell.

1.2 Molecular Chaperones

The term molecular chaperone has recently been introduced to define molecules that assist the process of folding and assembly of polypeptide chains without becoming part of the mature protein (Ellis 1987). Many of the known molecular chaperones are heat shock proteins (HSPs), some of which are expressed not only after stress induction but also in a constitutive fashion (reviewed in Lindquist 1986).

---

[1] Institut für Biochemie I, Universität Heidelberg, Im Neuenheimer Feld 328,
69120 Heidelberg, FRG.

44. Colloquium Mosbach 1993
Glyco- and Cellbiology

Each cellular compartment where protein folding occurs seems to possess its own set of molecular chaperones. In the endoplasmic reticulum (ER), the representative of the HSP70 family is BiP (Immunoglobulin Heavy Chain Binding Protein; Haas and Wabl 1983; Munro and Pelham 1986; Haas and Meo 1988), which is identical to GRP78, that was originally identified as one of two glucose regulated proteins (along with GRP94) present in fibroplasts whose synthesis is increased when cells are starved of glucose (Shiu et al. 1977). GRP94, also known as ERp99 (Mazarella and Green 1987), endoplasmin (Macer and Koch 1988), or gp96 (Srivastava et al. 1987), is an ER resident that belongs to the HSP90 family (Sorger and Pelham 1987). The mode of function of these proteins, however, is not yet understood.

BiP was originally discovered on grounds of its property to bind to immunoglobulin (Ig) heavy (H) chains in cells that do not express Ig light (L) chains (Haas and Wabl 1983). Usually, Ig H chains are not exported unless as part of a complete antibody molecule. However, removal of the CH1 domain leads to secretion of H chains even in the absence of L chain pairing and also abrogates normal BiP binding (Hendershot et al. 1987). This result demonstrates that the CH1 domain is essential for stable interaction of Ig H chains with BiP.

Subsequently, BiP was also found to interact stably with a variety of other polypeptide chains. A property common to all of these polypeptides may be their improper conformation, whereby the respective incorrect folding could be due to different causes: mutations in the primary structure (e.g., mutant viral hemagglutinin; Gething et al. 1986), incorrect disulfide bond formation or incorrect glycosylation (e.g., prolactin and yeast invertase, respectively; Kassenbrock et al. 1988). In this context, normal subunits of proteins unable to assemble could also be regarded as improperly folded structures (e.g., Ig H chains in the absence of L chain expression; Haas and Wabl 1983). On the other hand, BiP transiently interacts with polypeptide subunits which either must oligomerize (e.g., VSV G protein; Machamer et al. 1990) or assemble with an appropriate partner chain (e.g., Ig H chains when L chains are coexpressed; mouse muscle nicotinic receptor; Bole et al. 1986; Blount and Merlie 1991) in order to obtain their mature structure. These findings support the notion that BiP acts as part of a quality control system in the ER allowing aberrantly folded structures to correctly refold in the ER and, additionally, is also involved in the process of normal subunit assembly of multimeric proteins (Pelham 1989; Haas 1991).

Munro and Pelham (1986) demonstrated that the addition of ATP to an isolated BiP/ligand complex led to ATP hydrolysis and complex dissociation; ATP binding and ATPase activity being a general feature of all members of the HSP70 family. In vitro experiments revealed that the presence of peptides indeed stimulates the ATPase activity of BiP, presumably through their interaction with the protein binding site of BiP (Flynn et al. 1989). Further studies have shown that sequences of a more hydrophobic nature preferentially interact with BiP and that the critical size of a peptide to stimulate the ATPase activity of BiP was seven to eight amino acids in length (Flynn et al. 1991). Very recently, it was demonstrated that one of the peptides used by Rothman's group can indeed compete for the binding of BiP to Ig H chains (Gaut and Hendershot 1993). Although it is now well established that BiP can bind to polypeptide chains, and can release them by hydrolyzing ATP, it still remains puzzling how BiP proceeds in vivo to support the process of folding and assembly.

1.3 Biosynthesis of Antibody Molecules

Secretory glycoproteins, like antibodies, are translocated into the lumen of the endoplasmic reticulum (ER) of eukaryotic cells. The majority of the protein's molecular maturation steps occurs in this compartment. Catalyzed by the enzyme protein disulfide isomerase, disulfide bonds are formed or rearranged on the newly synthesized polypeptide chain, the first glycosylation steps occur, and subunits have probably assembled completely before the molecules reach the Golgi apparatus (Bergmann and Kuehl 1979; Bergmann et al. 1981; Roth and Pierce 1987).

In order to elucidate the role of BiP and other molecular chaperones in subunit assembly of oligomeric proteins, we decided to investigate the process of antibody formation. Advantages of using this system are (1) that antibodies are usually produced in large amounts by hybridoma cells, and (2) they are composed of different subunits (see Fig. 1). This allows us to investigate the fate of individual chains not only during the assembly process but also in the absence of partner chain expression.

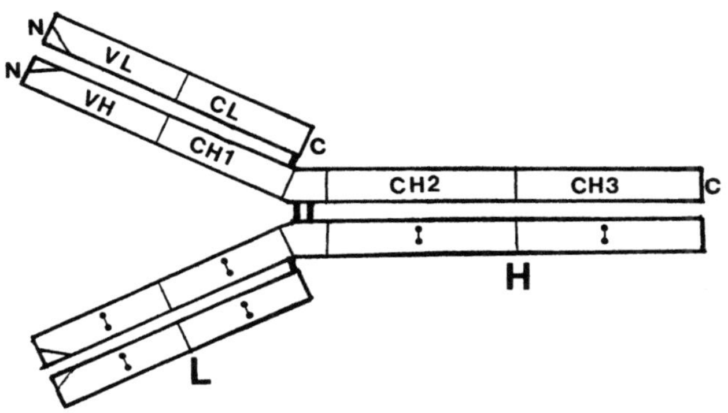

Fig. 1. The structure of antibody molecules. Antibody molecules are generally composed of two identical immunoglobulin. (Ig) heavy (*H*) and two identical light (*L*) chains which themselves are subdivided into different domains of about 110 amino acids in length, all of them exhibiting a similar three-dimensional structure (Ig fold) stabilized by a single internal disulfide bridge. H chains of an IgG molecule consist of three constant domains (*CH1–CH3*) and possess one variable domain (*VH*) at the amino terminus. In addition to a variable domain (*VL*), L chains bear one constant domain (*CL*) only. Chain pairing between H and L occurs between the V domains, whereby the antigen binding site is formed, and between *CL* and the first constant domain of the H chain (*CH1*) where also the interchain S-S bond is formed. Cysteines located in the region between *CH1* and *CH2* (hinge region) are involved in the formation of disulfide bridges between the two H chains. Despite the fact that the complete molecule is a tetramer, individual portions of the protein can be regarded as dimers

2 Interaction of BiP with Ig Chains

2.1 BiP and Immunoglobulin Heavy Chains

In the absence of L chain expression, BiP is complexed to normal H chains for a long period of time before the H chains are degraded. For mouse γ_3 H chains, we have determined the half-life of the complex to be approximately 2 h (Schröder, Kaloff, and Haas in prep.). The length of time BiP is bound to H chains, however, does not only depend on the isotype of the H chain (H chains of the μ isotype, for example, are degraded with a half-life of about 50 min; Sitia et al. 1987) but also on the availability of an L chain appropriate for pairing. In the presence of L chains, the same H chain is also bound to BiP but for a much shorter period of time (Bole et al. 1986), suggesting that H chains lose BiP binding as a consequence of L chain association. This view is strongly supported by the results of an in vivo experiment demonstrating that BiP-bound H chains were still capable of participating in the formation of complete antibody molecules (Hendershot 1990).

Molecular chaperones do not bind to correctly folded mature structures. Indeed, Ig H chains cannot be regarded as mature structures, since they represent incompletely assembled antibody molecules. Nevertheless, BiP-bound H chains can bind to protein A (Haas and Wabl 1983), which usually interacts with the folded CH2 and CH3 domains of complete IgG molecules (Deisenhofer et al. 1978). Thus, in BiP-bound H chains, these two domains may already exhibit the three-dimensional structure of a mature antibody molecule. This assumption is strongly supported by our recent finding that – in the absence of L chains – BiP does not only bind to H chain monomers but also to covalently linked H chain dimers (these analyses were performed under conditions not allowing chain oxidation to take place after lysis of the cells; Kaloff and Haas in prep.).

Sensitivity towards protease treatment is another means to analyze the folding state of a polypeptide chain. Using this approach, Neupert's group showed that a mitochondrial polypeptide became protease-resistant only when released from HSP60, indicating that the chaperone-bound molecule had not yet reached its mature conformation (Ostermann et al. 1989). In a similar type of experiment, we compared the protease sensitivity of H chains that were bound to or released from BiP. As shown in Fig. 2, BiP was released from H chains by the addition of MgATP. Interestingly, the amount of H chains isolated by Protein A Sepharose was not altered upon ATP treatment of the cell lysate, suggesting that dissociation of the BiP/H chain complex had not grossly affected the conformation of the CH2 and CH3 domains of the H chains. However, BiP-bound H chains were more sensitive towards protease treatment when compared to H chains released from BiP, indicating that a conformational change had taken place during dissociation of the complex (Fig. 2). Taken together, these results are in line with the notion that unassembled H chains exhibit different folding states along the polypeptide chain. Whereas the CH1 domain (which is presumably the portion of the molecule involved in BiP binding; see above) and possible also the V domain is not yet folded, the remaining portion has already assumed the conformation it will have in the mature antibody molecule. In conclusion, BiP binding to H chains

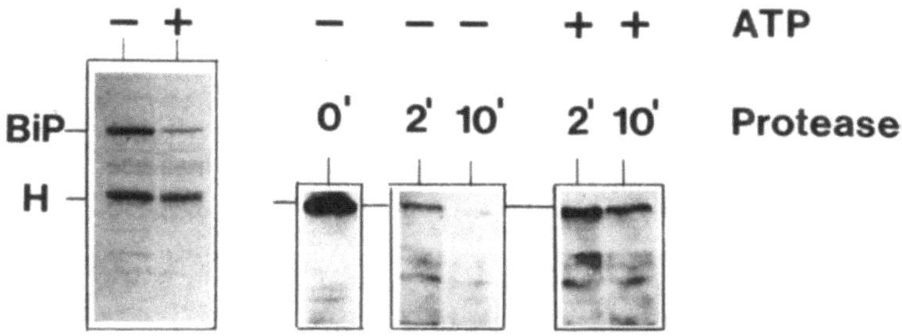

Fig. 2. Protease sensitivity of H chains. Lysates were prepared from H chain-producing cells (10^7 cells/ml) and treated (+) or not (–) with 5 mM MgATP (*ATP*). Thereafter, H chains were directly precipitated by Protein A Sepharose. The isolated material was analyzed by SDS-PAGE and visualized by Coomassie staining (*left panel*). Aliquots of the same lysates were incubated with Proteinase K (60 mg/ml, 4 °C, protease) for indicated time periods, after which the samples were boiled in reducing SDS-PAGE sample buffer and analyzed by Western blotting using a biotinylated goat antiserum directed against the corresponding mouse H chain isotype to reveal H chain-specific signals (*right panel*)

may serve to arrest that portion of the H chain in an assembly-competent conformation that later will be involved in L chain pairing.

2.2 BiP and Immunoglobulin Light Chains

BiP not only interacts with Ig H chains, but also binds to the other subunit of antibody molecules. The complex between BiP and L chains is formed either shortly after of in the course of L chain synthesis (Knittler and Haas 1992). Most of the L chains do not require H chain assembly to be exported from the ER. As expected for polypeptides that are secreted, the half-life of a complex between BiP and L chains turned out to be very short (2 min). In contrast, L chains that are not capable of being exported in the absence of H chain expression are dissociated from BiP only after a longer period of time (in the case analyzed, the half-life was 50 min). Interestingly, the kinetics of complex dissociation and L chain degradation were identical. It was therefore important to determine whether BiP was released from the complex before chain degradation occured. That this was indeed the case could be deduced from calculations based on both the determination of the percentage of total labeled BiP complexed to L chains and the half-life of BiP. Moreover, similar calculations performed on the basis of the data obtained from the L chain secreting cell line made it clear that BiP, once dissociated from its ligand, is recycled (Knittler and Haas 1992). Analyses by gel filtration chromatography showed that the complex of BiP and L chains was recovered from fractions corresponding to a molecular size of approximately 100–120 kDa, indicating that the complex contained a molar ratio of both components. This was confirmed by immunoprecipitation of L chains from a lysate which was previously treated with a reducible crosslinker. Under nonreducing conditions, the ATP-sensitive crosslink product migrated with an apparent molecular weight of 110 kDa in SDS-gels. BiP and L chains were released from this band when reanalyzed under reducing

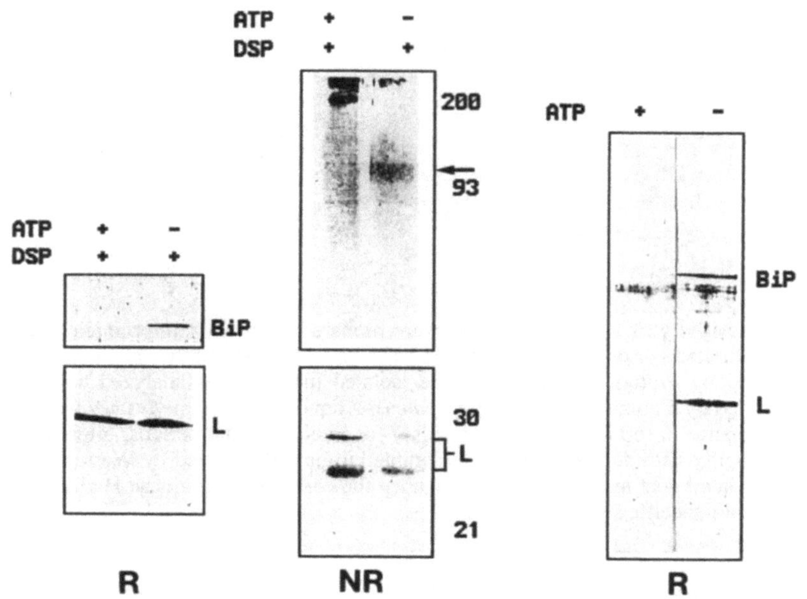

Fig. 3. BiP/L chain complexes contain a molar ratio of both proteins L chain-producing cells were lysed in the presence (+) or absence (–) of 5 mM MgATP (*ATP*). The lysates were then treated (+) or not (–) with 500 µM of a thio-reducible crosslinker (*DSP*). L chains were immunoprecipitated and the isolated material was separated on SDS gels under reducing (*R*, *left panel*) or nonreducing (*NR*, *middle*) conditions before being transferred onto nitrocellulose. In a parallel experiment, the regions corresponding to the apparent molecular weight of the ATP-sensitive crosslinking product were cut out from both lanes of the nonreducing gel (*arrow*) and reanalyzed under reducing conditions (*right panel*). The nonreducing Western blot (*middle*) was developed with anti-L chain antiserum only, whereas the reducing ones (*left and right panels*) were developed with both anti-L chain and anti-BiP reagents

conditions (Fig. 3). Moreover, when a similar experiment was repeated with radioactive labeled proteins, BiP and L chains were the only labeled proteins contained in the crosslink product. Considering the half-lives of BiP and L chains, it was possible to calculate and confirm the molar composition of labeled proteins contained within the complex after microdensitometric quantification of the signals individually given by L chains and BiP, respectively. Though these results do not exclude the existence of larger complexes containing additional components in the ER, which may have disintegrated during the isolation procedure, they demonstrate that one molecule of L chains interacts with one molecule of BiP.

3 GRP94 and Immunoglobulin Chains

3.1 The Observation of Additional Proteins that Coisolate with BiP

In the course of the experiments performed to determine the half-life of BiP (Knittler and Haas, 1992), we by chance observed the coisolation of additional long-lived proteins with BiP and/or Ig chains. As shown below, these findings were due to the special experimental conditions used to analyze a protein that has a relatively long half-life.

Cells were pulse-labeled for 60 min and thereafter kept under normal cell culture conditions. At different time points of chase, identical culture volumes were taken and all cells contained in these were solubilized. Both the expected time-dependent change in the overall pattern of labeled proteins and the enrichment of long-lived labeled molecules were rendered visible by analyzing either equal amounts of radioactively labeled material or equal volumes of the corresponding lysates (Fig. 4 a). Note that, in addition to BiP (78 kDa), several other long-lived proteins were enriched after 96 h of chase. The analysis of BiP immunoisolated from identical amounts of lysates prepared at the different time points of chase revealed an unexpected picture. Two additional proteins with apparent molecular weights of approximately 94 and 170 kDa, respectively, were also evident, but only at the later time points of chase (Fig. 4 b). The same result was obtained when H chains were isolated in a pulse chase experiment performed with H chain producing cells (Fig. 4 c). Since Protein A-Sepharose alone was used to directly isolate the H chains, this result suggested that coisolation of the additional proteins was not due to crossreactivity of the anti-BiP antibody used in the first experiment. In contrast, Protein A-Sepharose did not precipitate any material from lysates that contained a comparable amount of labeled protein but prepared from cells that did not produce H chains (Fig. 4 c). Furthermore, similar patterns were obtained after the analysis of L chain-producing cells when the chains were isolated. Thus, the appearance of these additional proteins was dependent on the specific isolation of BiP or BiP/Ig chain complexes, indicating that the two proteins are complexed to Ig chains and/or to BiP. From these analyses, it could not be determined whether the complexes were formed by direct or indirect interaction of the corresponding proteins.

However, why did these proteins only appear after a longer chase period? The only valid interpretation was to assume that the steady-state levels of the complex had increased during the chase period because the cells had divided in culture. Therefore, at day 2 or 4 of chase, the precipitations were presumably performed from lysates containing more protein in total. If this explanation was true, the cells probably only contain low steady levels of the complex. In order to confirm this, we repeated the experiment, but additionally also precipitated BiP from ten times more lysate prepared directly after the pulse. Because of the high amount of radiolabeled proteins present in this lysate (corresponding to 5×10^7 cpm), the nonspecific background of the precipitation was too high to detect coprecipitation of a 170-kDa protein (Fig. 4 d). However, when analyzing a low exposure of the gel, a band comigrating with the 94-kDa protein became visible, among other proteins, indicating that our original assumption was correct. From our results we conclude that two additional proteins are

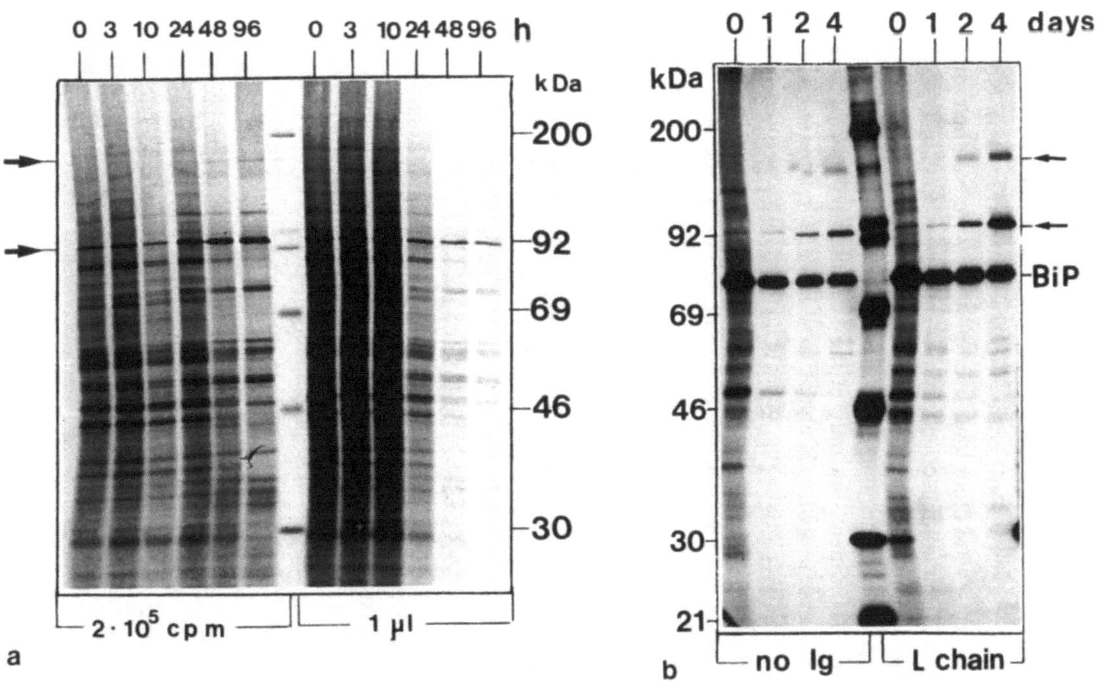

Fig. 4 a, b. Two unidentified proteins are complexed to BiP/Ig chains. **a** Hybridoma cells were labeled (1 h) with ^{35}S-methionine (10 MBq/10^6 cells in 500 µl methionine-free label medium) and chased under normal cell culture conditions (5 x 10^4 cells/ml) in the presence of an excess of unlabeled methionine. At the time points indicated, aliquots of the cultures were taken and the cells contained within were solubilized in a defined volume of lysis buffer. Note that, since the cells continued to grow, and divided in culture, the different lysates prepared in this way did not contain the same amount of total cellular proteins. After determination of TCA-precipitated radioactivity, either the same amount of incorporated radioactivity (*left*) or equal volumes of the lysates (*right*) were applied onto an SDS-gel and the radiolabeled proteins were visualized by fluorography. The *arrows* indicate the two long-lived proteins later observed to coprecipitate with BiP or BiP/Ig chain complex. **b** Two parent hybridoma lines, producing either *no Ig* or *Ig L* chains, were pulse labeled and chased as described in **a**. Immunoisolation of BiP was performed from aliquots of the different lysates (corresponding to the volume of lysate prepared directly after the pulse that contained 5 x 10^6 cpm) and isolated material was analyzed by SDS-PAGE and fluorography. Migrating at their apparent molecular weights, the 94- and 170-kDa proteins (*arrows*) correspond to the two long-lived proteins indicated in **a**

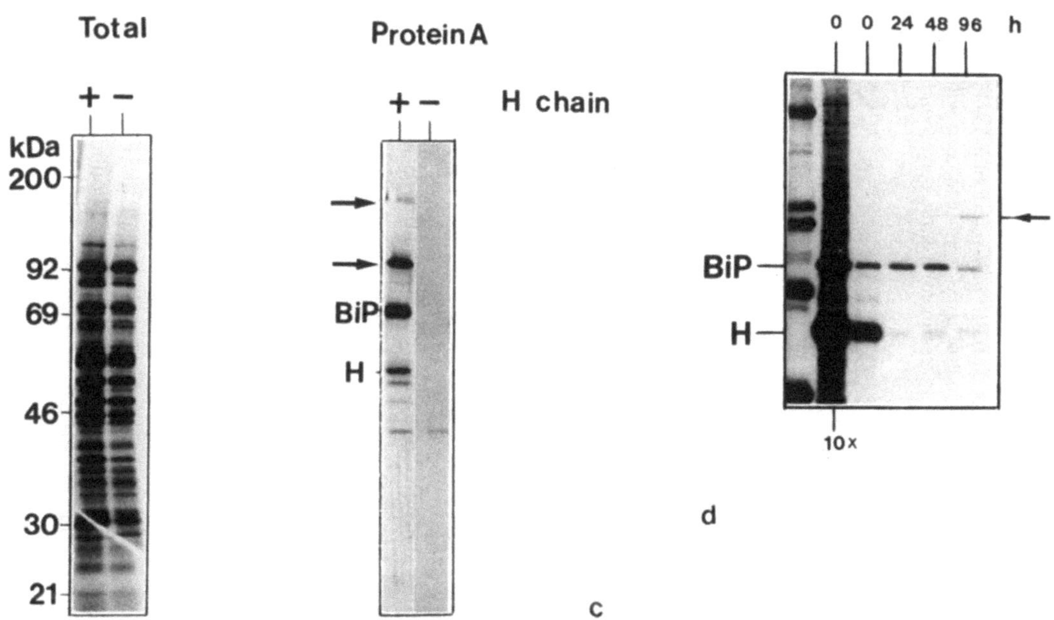

Fig. 4 c, d, Two parent hybridoma lines producing either Ig H chains (+) or no Ig (–) were pulse labeled and chased as described in the legend to **a**. From each of the lysates prepared at day 4 of chase (*Total*), Protein A Sepharose-binding material was analyzed (*Protein A*). Note that these precipitations were performed in the absence of antiserum. **d** A H chain-producing hybridoma line was pulse labeled and chased as described in **a**. Protein A Sepharose was used to directly precipitate H chains from aliquots of the lysates prepared as described in **b** (*0–96 h*). H chains were also precipitated from ten times (*10x*) more lysate prepared from cells isolated directly after the pulse

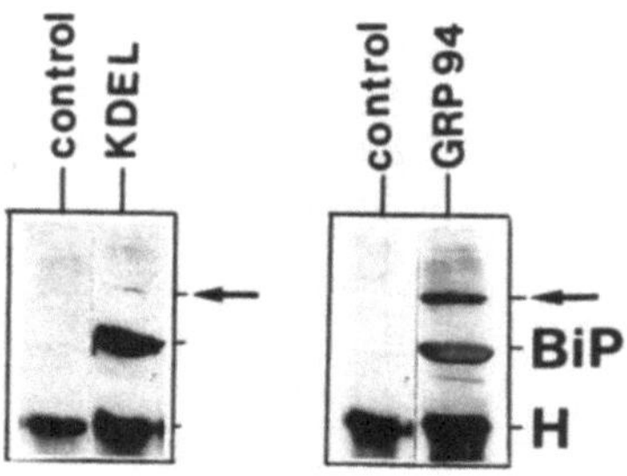

Fig. 5. Identification of GRP94. A lysate prepared from H chain-producing hybridoma cells was used to isolate Ig H chains. After washing, the Protein A Sepharose-bound material was separated on an SDS-gel and transferred onto nitrocellulose. The Western blots were treated either with a monoclonal rat antibody reacting with the carboxy terminal tetrapeptide common to BiP and GRP94 (*KDEL*) and developed with a biotinylated monoclonal mouse antibody directed against rat κ L chains or with a rabbit anti-GRP94 antiserum raised against the C-terminal 16 amino acids of mouse GRP94 (which also reacts with BiP in Western blots) and developed with biotinylated goat anti-rabbit antiserum. The control lanes were incubated with irrelevant rat monoclonal antibody or rabbit antiserum prior to treatment with the corresponding second antibody reagent (*control*)

coisolated by immunoprecipitation of BiP and by isolation of Ig chains complexed to BiP, respectively.

3.2 Identification of the 94-kDa Protein as GRP94

Because of its molecular size, we speculated that one of the proteins coisolating with BiP or BiP/Ig chain complexes was GRP94. This was first confirmed by immunoprecipitations using an antiserum directed against the amino terminal 16 amino acids of mouse GRP94 (a kind gift of Michael Green, St. Louis, USA). When the electrophoretic mobilities were compared, both immunoisolated GRP94 and the 94-kDa protein coisolated with BiP/Ig chains migrated at identical positions (see Fig. 6). Furthermore, the analysis of precipitated H chains in Western blots revealed the presence of a GRP94-sized protein band when anti-GRP94 antiserum or an antibody reacting with the carboxy terminal tetrapeptide common to BiP and GRP94 (kind gift of Geoffrey Butcher, Cambridge, UK) was used for immunostaining (Fig. 5). This demonstrated that one of the proteins coisolating with BiP/Ig chain complex is GRP94.

3.3 A Supramolecular Chaperone Complex in the ER?

There are several possibilities to explain the coprecipitation of GRP94 with BiP and Ig chains: (1) GRP94 is bound to BiP and BiP in turns is bound to H chains; (2) GRP94 is bound to H chains and the same or different H chains are bound to BiP; (3) GRP94 is bound to a third protein which in turn is bound either to BiP or to Ig chains. In order to analyze the first possibility, we dissociated the BiP/ligand complex prior to immunoprecipitation of Ig chains. As clearly seen from the result of a Western blot

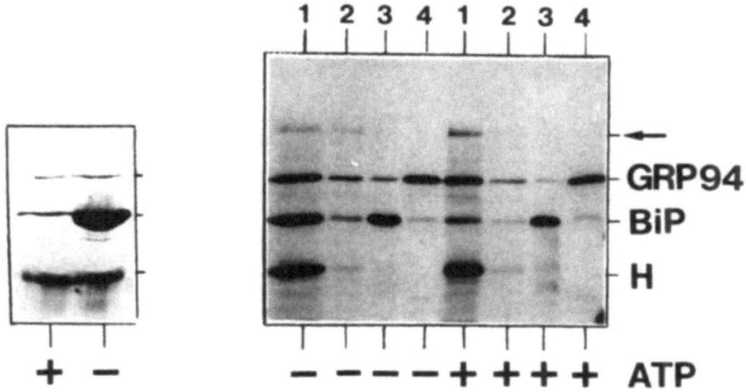

Fig. 6. GRP94 interaction with H chains is not sensitive to ATP. The *left panel* shows the result of a Western blot performed as described in the legend to Fig. 5 except that the lysates were treated (+) or not (–) with 5 mM MgATP prior to H chain precipitation. The blot was developed using antiserum to the C-terminal portion of GRP94. For the experiment shown on the *right*, H chain producing cells were pulse labeled and chased for 96 h prior to lysis which was performed in the presence (+) or absence (–) of 5 mM MgATP. Precipitations were done in the following way: two consecutive Protein A Sepharose incubations were to isolate H chains (*lane 1*) and to accomplish depletion of remaining Protein A-binding material (*lane 2*). Subsequently, the remaining lysate was split into two parts to isolate either BiP (*lane 3*) or GRP94 (*lane 4*) by using a monoclonal anti-BiP antibody (Bole et al. 1986) or anti-GRP94 antiserum raised against the amino terminal 16 amino acids, respectively. In both cases, Protein A Sepharose was used to precipitate the antibodies

experiment, BiP was released from the complex in the presence of MgATP (Fig. 6). In contrast, most of the GRP94 still coisolated with H chains. In order to analyze the ATP dependency of the interaction between H chains and the 170-kDa long-lived protein, we performed a similar experiment using biosynthetically labeled cells lysed at day 4 of chase in the presence or absence of ATP. Figure 6 (lane 1, ± ATP) compares the ATP dependency of the coisolation of BiP, GRP94, and the 170-kDa proteins with H chains. Whereas BiP dissociated from the complex, both GRP94 and the 170-kDa protein were still coprecipitated with H chains. Interestingly, a small amount of BiP was also coisolated with GRP94 and vice versa irrespective of whether or not the lysates were treated with ATP (Fig. 6, lanes 3 and 4, respectively). Thus, GRP94 may also directly interact with BiP in the absence of any ligand. Independent of the latter observation, these results show that the coisolation of GRP94 and the 170-kDa component with Ig chains was not due to a direct or indirect interaction of these two proteins with BiP, but to a direct or indirect interaction with Ig chains instead. Moreover, both GRP94 and the 170-kDa protein either do not exhibit an ATPase activity on their own or do not use this property to dissociate from their ligands. In order to further investigate the molecular composition of the complex, we depleted the lysates of those complexes that contained BiP by consecutive immunoprecipitations. Astonishingly, as much as three consecutive precipitations were not sufficient to completely remove all of the labeled BiP from the lysate (Fig. 7). However, the BiP signal decreased with each precipitation step, indicating that most of the protein had been isolated. Nevertheless, it was still possible to isolate H chains from the remaining lysate by the use of Protein A Sepharose. Moreover, GRP94 and the 170-kDa protein still

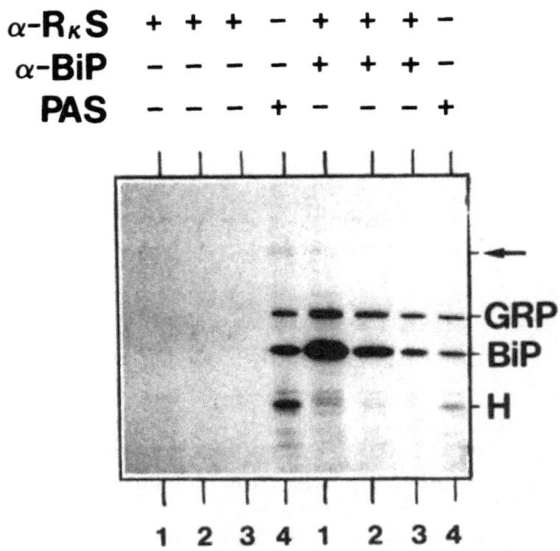

Fig. 7. Depletion of BiP-containing complexes. H chain-producing hybridoma cells were pulse labeled and chased for 96 h as described in the legend to Fig. 4a. From lysates corresponding to approx. 2 x 10^6 cells, consecutive precipitations (*1–4*) were performed as indicated. The precipitating reagents used were: monoclonal antibody to rat k L chains coupled to Sepharose (a-RkS), monoclonal anti-BiP antibody (Bole et al. 1986), (*a-BiP*), and Protein A Sepharose (PAS). The *arrow* indicates the 170-kDa protein

coisolated with the H chains (Fig. 7, lane 4). When microdensitometric quantification was used to compare the amount of labeled protein isolated with H chains, with or without prior depletion of BiP-containing populations, we noticed that the molar composition of the complex had changed. Whereas the ratio of labeled H chains and labeled GRP94 remained constant (though the amount of both had diminished), the signal for BiP was disproportionately lower. We interpret these results in the following way: the anti-BiP antibody does not react equally well with all forms of BiP and reacts particularly poorly with BiP involved in a complex which contains H chains, GRP94, and the 170-kDa component. This means that besides binary BiP/H chain complexes (which are removed by anti-BiP precipitations), a supramolecular complex may exist which additionally contains GRP94 and the 170-kDa protein. We are, however, aware of the fact that the existence of such complexes remains to be formally proven.

4 Final Conclusions

In the experiments described above, we have shown that unassembled Ig chains not only bind to BiP but also to GRP94 and a still unidentified protein of 170-kDa apparent molecular weight. It still remains unclear whether a given H chain molecule is simultaneously complexed to BiP, GRP94, and the 170-kDa component, respectively. Alternatively, individual complexes may exist, each involving two components only.

A third possibility – and the most difficult to resolve – would be the existence of various complexes each one exhibiting a different molecular composition. Very recently, Melnick and coworkers also reported on GRP94 interaction with Ig chains, claiming that GRP94 forms ternary complexes including both BiP and Ig chains (Melnick et al. 1992). From the data presented, however, it is not clear whether the 170-kDa protein was also contained in the crosslinked complex. For the following reasons, we believe that only a minor population of H chains is complexed with GRP94 and the 170-kDa protein: (1) when H chains are precipitated from unlabeled lasates and analyzed in Coomassie stained gels, only BiP is detected (Fig. 2). In some gels, a faint band of GRP94 is sometimes observed, the 170-kDa protein, however, is never seen. (2) crosslinking of the lysates prior to immunoprecipitation of L chains reveals one major crosslink product that corresponds to a binary complex of BiP and L chains (Fig. 3); (3) when BiP and GRP94 coisolated with H chains are stained with the antibody to KDEL in Western blots, the amount of GRP94 appears to be drastically lower as compared to the amount of BiP (Fig. 5). If these findings reflect the real distribution of complexes present in the ER, it would mean that a small portion of Ig chains are complexed with GRP94, the 170-kDa protein, and possibly BiP, whereas the major fraction of the Ig chains form a binary complex with BiP. In terms of interaction kinetics, one could speculate that the interaction of these additional proteins occurs in the beginning or at the end of the period BiP interacts with the folding polypeptide chain. It is therefore attractive to postulate that GRP94 (and/or the 170-kDa component) may be involved in the initiation of the chaperone/polypeptide chain complex formation. Alternatively, the other chaperones may be required to allow dissociation of the BiP/ligand complex, an event which, to our understanding, must occur in a controlled fashion. These two functions, however, are not mutually exclusive.

Acknowledgments. We wish to thank Drs. Raul Torres and L. Hendershot for critical reading of the manuscript. Drs. Michael Green and Geoffrey Butcher are gratefully acknowledged for their generous gift of anti-GRP94 antisera and monoclonal anti-KDEL antibody, respectively. M.R. Knittler receives a fellowship from the Boehringer/Ingelheim Fonds. This work was supported by the Deutsche Forschungsgemeinschaft through SFB243.

References

Anson ML (1945) Adv Protein Chem 2:361–384
Bergmann LW, Kuehl WM (1979) J Biol Chem 254:5690–5694
Bergmann LW, Harris E, Kuehl WM (1981) J Biol Chem 256:701–706
Blount P, Merlie JP (1991) J Cell Biol 113:1125–1132
Bole DG, Hendershot LM, Kearney JF (1986) J Cell Biol 102:1558–1566
Burnet FM (1941) The production of antibodies, 1st edn Macmillan, Melbourne
Deisenhofer J, Jones TA, Huber R, Sjödahl J, Sjöquist J (1978) Hoppe Seyler's Z Physiol Chem 359:975–985
Ellis J (1987) Nature 328:378–379
Flynn GC, Chappel TG, Rothman JE (1989) Science 245:385–390
Flynn GC, Pohl J, Flocco MT, Rothman JE (1991) Nature 343:726–730
Gaut JR, Hendershot LM (1993) J Biol Chem 268:7248–7255
Gething MJ, McCammon K, Sambrook J (1986) Cell 46:939–950

Haas IG, Wabl MR (1983) Nature 306:387–389
Haas IG, Meo T (1988) Proc Natl Acad Sci USA 85:2250–2254
Haas IG (1991) Curr Top Microbiol Immunol 167:71–82
Haber E (1964) Proc Natl Acad Sci USA 52:1009
Hendershot LM, Bole D, Köhler G, Kearney JF (1987) J Cell Biol 104:761–767
Hendershot LM (1990) J Cell Biol 111:829–837
Jaenicke R (1988) 39th Colloquium Mosbach. Protein structure and protein engeneering. Springer, Berlin Heidelberg New York
Kassenbrock CK, Garcia PD, Walter P, Kelly RB (1988) Nature 333:90–93
Knittler MR, Haas IG (1992) EMBO J 11:1573–1581
Landsteiner K (1945) The specificity of serological reactions. Harvard University Press, Boston
Lindquist S (1986) Annu Rev Biochem 55:1151–1191
Macer DPJ, Koch GLE (1988) J Cell Sci 92:61–70
Machamer CE, Doms RW, Bole DE, Helenius A, Rose JK (1990) J Biol Chem 265:6879–6883
Mazzarella RA, Green M (1987) J Biol Chem 262:8875–8883
Melnick J, Aviel S, Argon Y (1992) J Biol Chem 267:21303–21306
Munro S, Pelham HRB (1986) Cell 46:291–300
Ostermann J, Horwich A, Neupert W, Hartl FU (1989) Nature 341:125–130
Pelham HRB (1989) Annu Rev Cell Biol 5:1–23
Roth RA, Pierce SB (1987) Biochemistry 26:4179–4182
Shiu RPC, Pouyssegur J, Pastan I (1977) Proc Natl Acad Sci USA 74:3840–3844
Sitia R, Neuberger M, Milstein C (1987) EMBO J 6:3969–3977
Sorger PK, Pelham HRB (1987) J Mol Biol 194:341–344
Srivastava PK, Chen Y-T, Old LJ (1987) Proc Natl Acad Sci USA 84:3807–3811
Whitney PL, Tanford C (1965) Proc Natl Acad Sci USA 53:524

Function of Molecular Chaperones in Protein Folding

F. U. Hartl[1]

1 Introduction

How a newly synthesized polypeptide chain acquires its specific three-dimensional structure is a fundamental problem in biology. Only in recent years, however, have cell biologists and physical biochemists turned their attention to the process of protein folding as it occurs in vivo. As a result, the long-held view that proteins in the cell fold in a largely spontaneous process has been called into question.

2 The Molecular Chaperone Concept

Although the basic principle that the primary sequence of a protein is sufficient to specify its three-dimensional structure remains unchallenged (Anfinsen 1973; Creighton 1990; Jaenicke 1991), it has become clear that protein folding in the cell depends on helper proteins and requires the input of metabolic energy (Gething and Sambrook 1992; Hendrick and Hartl 1993). A multitude of such proteins, collectively classified as molecular chaperones (Ellis 1987), have been discovered. Many of them are heat-shock or stress proteins, but have essential functions also under normal cellular conditions. They recognize unfolded or partially denatured proteins and do not interact with proteins in their native state. Recent studies indicate that the predominant role of chaperones is to prevent the incorrect intra- and intermolecular association of polypeptide chains which results in their aggregation. The physiological importance of this can easily be rationalized: rapid spontaneous re-folding of purified proteins in vitro requires the presence of a complete polypeptide or at least a folding domain (Fischer and Schmid 1990). Under cellular conditions, however, proteins emerge from ribosomes (or from the *trans*-side of a membrane during translocation) as unfolded chains which are restricted from forming stable tertiary structure. Unfolded proteins or partially folded intermediates expose hydrophobic surfaces and thus have the tendency to aggregate given the high cellular concentration of folding polypeptides (Creighton 1990; Christensen and Pain 1991; Jaenicke 1991). It is therefore believed that productive folding in vivo first depends on the prevention of incorrect folding by molecular chaperones at a stage when the polypeptide chain is still nascent, i.e., prior to its release from the ribosome.

---

[1] Rockefeller Reserach Laboratories, Sloan-Kettering Institute, 1275 York Avenue, New York, NY, 10021, USA.

44. Colloquium Mosbach 1993
Glyco- and Cellbiology

3 Sequential Action of Molecular Chaperones in Protein Folding

The function of preventing incorrect folding during synthesis or membrane translocation of a protein is carried out by chaperones acting as *monomers* or *dimers* such as the constitutively expressed members of the Hsp70 family of stress proteins. Once a sufficient length of polypeptide chain is available for productive folding (or upon completion of translation/translocation in the case of single-domain proteins), a second principle of chaperone action can take over: *olgigomeric chaperones*, such as the members of the Hsp60 (chaperonin) family, mediate folding to the native state by providing an environment that protects the polypeptide chain from unproductive interactions during the folding process.

The functional cooperation of different types of molecular chaperones with different specificities in a sequential folding reaction has become apparent by studying mitochondrial protein import (for a review see Hartl et al. 1992). In order to grow and divide, mitochondria take up most of their proteins from the cytosol as precursors that typically contain amino-terminal targeting sequences. Precursor proteins are translocated across the two mitochondrial membranes in a rather extended conformation. This occurs at translocation contact sites where the two mitochondrial membranes are in close proximity. Figure 1 shows a model for the pathway of chaperone-mediated folding in mitochondria.

The Hsp70 in the mitochondrial matrix interacts with the incoming, extended polypeptide chains (Neupert et al. 1900; Kang et al. 1990; Scherer et al. 1990). This interaction may resemble that of cytosolic Hsp70 with nascent polypeptide chains emerging from ribosomes (Beckman et al. 1990). It is required for efficient translocation of proteins but, at least in most cases, appears not to be sufficient for their subsequent folding and assembly. Mitochondrial Hsp60 receives the newly imported poly-

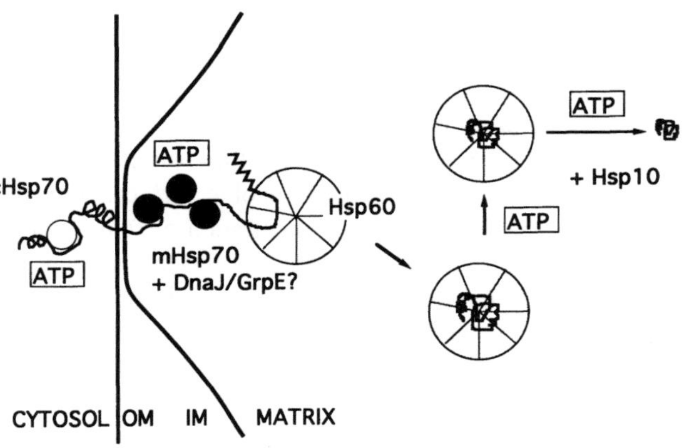

Fig. 1. Model for the sequential action of Hsp70 and Hsp60 in the folding of proteins imported into mitochondria. Mitochondria are likely to contain homologues of *E. coli* DnaJ and GrpE which have not yet been identified. The amino-terminal targeting sequence of the precursor protein (*zig-zag*) is cleaved during or after translocation by the mitochondrial processing enzyme (not shown). *cHsp70*, Cytosolic Hsp70; *mHsp70* mitochondrial Hsp70; *OM* outer membrane; *IM* inner membrane

peptide chains from Hsp70 and mediates their folding to the native state in an ATP-dependent process (Cheng et al. 1989; Ostermann et al. 1989; Manning-Krieg et al. 1991). In this reaction, Hsp60 requires the regulation by a further stress protein, Hsp10, which is the mitochondrial homologue of *E. coli* GroES (Hartman et al. 1992).

Using the Hsp70 and Hsp60 homologous stress proteins of *E. coli*, DnaK, and GroEL, we have recently reconstructed a folding reaction in which molecular chaperones of the Hsp70 and Hsp60 classes act sequentially (Langer et al. 1992a). It turned out that this process requires additional components, namely the heat-shock proteins DnaJ and GrpE, which regulate the function of the Hsp70 protein DnaK (Georgopulos et al. 1990). The overall reaction can be subdivided into the following three steps: (1) stabilization of an aggregation-prone folding intermediate by DnaK and DnaJ; (2) GrpE-dependent transfer of this intermediate to the chaperonin GroEL; (3) ATP- and GroES-dependent folding to the native state at GroEL. Our experimental approach was based on first unfolding a substrate protein in 6 M guanidinium-Cl and then diluting it into buffer solution containing the various chaperone proteins. We chose the monomeric mitochondrial protein rhodanese as the substrate for these studies because of its pronounced tendency to aggregate upon attempted refolding by dilution from denaturant in vitro (Tandon and Horowitz 1986). DnaK showed only a weak potential to prevent the aggregation of rhodanese. Only when DnaK was used at a high (10–20-fold) molar excess over rhodanese did we see a significant stabilizing effect (Langer et al. 1992a). It appeared possible, therefore, that DnaK required the cooperation of an additional chaperone protein. A good candidate for this function was the stress protein DnaJ, which was known to regulate the ATPase of DnaK (Liberek et al. 1991). Surprisingly, we found that DnaJ alone was quite efficient in interacting with unfolded rhodanese identifying this protein as a molecular chaperone. Furthermore, when combined with DnaK, there was a clear cooperative effect of DnaJ and DnaK in preventing protein aggregation (Langer et al. 1992a).

The mitochondrial model system suggests that Hsp70 (presumably in cooperation with a homologue of DnaJ) interacts first with the unfolded polypeptide chains entering the mitochondrial matrix before they reach Hsp60 (Fig. 1). It seemed reasonable to assume, therefore, that the sequence of interactions between folding proteins and molecular chaperones is guided by a differential substrate specificity of the chaperone components for increasingly folded polypeptides. To address this, we have compared the binding efficiencies of DnaK (Hsp70), DnaJ and GroEL towards three different substrates: reduced carboxymethylated alpha-lactalbumin (RCMLA) served as model for an extended polypeptide chain perhaps similar to a nascent chain early in translation. Casein, although stable in solution, exposes hydrophobic surfaces and has certain properties of compact folding intermediates. Rhodanese, upon dilution from denaturant, undergoes rapid collapse to a "molton glubule"-like intermediate. Only DnaK was able to recognize the extended RCMLA (Langer et al. 1992a). DnaJ and GroEL had similar binding properties in these assays. Both components interacted with casein and rhodanese, suggesting that they recognize partially folded intermediates. In collaboration with M. Wiedemann, we have recently found that DnaJ has the capacity to bind to nascent polypeptide chains very early in translation thus preventing the folding of newly synthesized proteins (Hendrick et al. 1993). Based on these observations, it appears likely that DnaJ recognizes peptide segments in unfolded polypeptides, presumably with a specificity different from that of Hsp70. Hsp70 and DnaJ (or DnaJ homologues) may generally cooperate in preventing incorrect folding of proteins during translation or membrane translocation.

We tested whether DnaK/DnaJ-stabilized rhodanese was productive for folding to the native state. When Mg-ATP was added to the protein complex, this did not result in folding of rhodanese to the active enzyme. Quite unexpectedly, rhodanese was even more tightly bound by DnaK and DnaJ when Mg-ATP was present. In contrast, DnaK alone releases its substrate upon ATP hydrolysis. Thus, DnaJ modifies the interaction of DnaK with an unfolded protein. It seemed possible that GrpE was necessary in addition to obtain the reactivation of rhodanese. GrpE is a small heat-shock

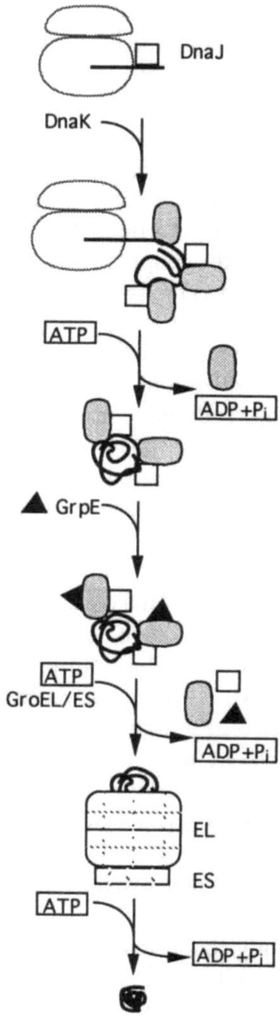

Fig. 2. Model for the sequential action of DnaK (Hsp70)/DnaJ/GrpE and GroEL/GroES in folding newly synthesized proteins in the cytosol of *E. coli*. The sequence of interactions of DnaJ and DnaK with the nascent polypeptide has not yet been determined

protein that is known to stimulate the ATPase activity of DnaK when DnaJ is present (Liberek et al. 1991). However, addition of GrpE to DnaK/DnaJ-stabilized rhodanese resulted in only a slow and very inefficient reactivation. Next, we tested whether the DnaK/DnaJ-bound rhodanese could be transferred to GroEL for folding. Again, addition of just GroEL and GroES was without effect, but when GrpE was also added, this resulted in efficient reactivation of rhodanese in a GroEL-, GroES-, and ATP-dependent reaction (see below).

The properties of this transfer reaction were analyzed in more detail. For rhodanese which was stably bound by DnaK and DnaJ in the presence of Mg-ATP, transfer to GroEL was strictly dependent on GrpE. However, in the absence of ATP, rhodanese was more loosely bound by DnaK/DnaJ and could be transferred to GroEL in an ATP-independent step. Recent experiments indeed indicate that not ATP-hydrolysis but rather the GrpE-catalyzed exchange of DnaK-bound ADP for ATP is required for the release of the DnaK/DnaJ-stabilized protein (our unpubl. observ.). The possibility of efficient transfer of a partially folded substrate protein from DnaK/DnaJ to GroEL also allowed us to test the ability of GroEL to carry out multiple rounds of rhodanese folding. GroEL was added in substoichiometric concentrations to a solution containing an excess of rhodanese stabilized by DnaK and DnaJ. Upon addition of GrpE, efficient reactivation was observed whereby one GroEL 14-mer complex was able to carry out four to six rounds of rhodanese folding (Langer et al. 1992a).

These reactions are summarized in a hypothetical model drawn for the folding of proteins in the bacterial cytosol (Fig. 2). The degree of coupling between the DnaK/DnaJ and the GroEL/ES systems is currently under investigation. Certain proteins may fold efficiently following their GrpE-dependent release from DnaK/DnaJ and may not have to interact with GroEL. For example, the folding of firefly luciferase can be mediated by DnaK/DnaJ/GrpE alone, at least in vitro (unpubl. observ. with B. Bukau). Interestingly, however, this option is not followed when GroEL is present and a luciferase folding intermediate is then transferred to the chaperonin upon its GrpE-catalyzed release from DnaK/DnaJ.

4 Double Ring Chaperonins Mediate Protein Folding

The paradigm of molecular chaperones that mediate folding to the native state in an ATP-dependent reaction are the members of the Hsp60 family, the so-called chaperonins (Hemmingsen et al. 1988). These proteins occur in the bacterial cytosol (GroEL of *E. coli*), in the matrix space of mitochondria (mitochondrial Hsp60) and in the stromal space of chloroplasts (rubisco binding protein) (for a review see Hendrick and Hartl 1993). The chaperonins are active as ~800 kDa oligomeric complexes composed of two stacked rings of seven 60-kDa subunits. They have a weak ATPase activity that requires functional regulation by a cochaperonin (GroES of *E. coli*; Hsp10 of mitochondria), a single ring of seven 10-kDa subunits that forms a complex with the chaperonin double ring (Viitanen et al. 1990; Martin et al. 1991; Langer et al. 1992b). The chaperonins are absolutely essential for cell growth under all conditions. Their function in protein folding has been recognized through a series of studies in vivo and with isolated organelles: (1) GroEL mutants of *E. coli* fail to assemble head structures of phage λ and T5 (Georgopoulos et al. 1973); (2) a complex found in the chloroplast stroma between the cytoplasmically synthesized large subunit of ribulose bisphosphate carboxylase and the "rubisco subunit binding protein" was shown to be an intermediate in the oligomeric assembly of this enzyme (Cannon et al. 1986; Hemmingsen et al. 1986); (3) evidence that protein assembly in general might require chaperonins came from the study of *mif4* mutant yeast which harbors a temperature-sensitive mitochondrial Hsp60. Although in these yeasts the import of proteins into the organelle occurs normally, newly imported proteins fail to assemble (Cheng et al. 1989, 1990). The kinetic and spatial separation of protein synthesis and folding/assembly in this experimental system has subsequently been utilized to demonstrate that the mitochondrial chaperonin supports even the folding of simple monomeric proteins, such as dihydrofolate reductase (DHFR) (Ostermann et al. 1989; Martin et al. 1992), which are well able to fold spontaneously in vitro. Since then, it has become apparent through numerous studies that the basic role of the chaperonins is indeed the folding of protein monomers, although a specific function in subunit association cannot be ruled out.

5 Reconstitution of Chaperonin Function

First insight into the complex molecular mechanism of chaperonin action has been obtained through the reconstitution of chaperonin function in vitro (Goloubinoff et al. 1989; Buchner et al. 1991; Martin et al. 1991). Most of these studies have been carried out with the Hsp60 of *E. coli*, GroEL. The GroEL oligomer has been shown to bind only one or two molecules of unfolded polypeptide via as yet undefined structural elements. Binding to the chaperonin is usually achieved by diluting unfolded polypeptide from strong denaturant into chaperonin-containing buffer solution. This prevents the aggregation of unfolded proteins such as rhodanese or citrate synthase. GroEL is then able to mediate the efficient refolding of these proteins to the native state in a reaction that is strictly dependent on ATP hydrolysis and GroES (Buchner et al. 1991; Martin et al. 1991; Mendoza et al. 1991). In contrast, other substrate proteins, such as DHFR, are independent of GroES and ATP hydrolysis for their GroEL-mediated folding (Martin et al. 1991; Viitanen et al. 1991). These proteins are characterized by a low tendency to aggregate under optimal refolding conditions in vitro. A common functional principle of chaperonin action explaining both types of reactions has yet to be established.

GroEL appears to bind its substrate protein in a loosely folded conformation resembling the "compact intermediate" or "molten globule"-like state (Martin et al. 1991) that may contain native secondary structure but lacks stable tertiary structure (Christensen and Pain 1991). In collaboration with W. Baumeister and coworkers, we have recently analyzed structural and functional properties of the interaction between GroEL, its cochaperonin, GroES, and substrate polypeptide. The stoichiometry of GroEL-14mer to GroES-7mer in the holo-chaperonin complex was found to be 1:1, indicating an asymmetrical structure for the functional chaperonin unit. Electron microscopic image analysis of GroEL-GroES complexes confirmed the asymmetrical shape of the holochaperonin. We found that GroES binds to one end-surface of the

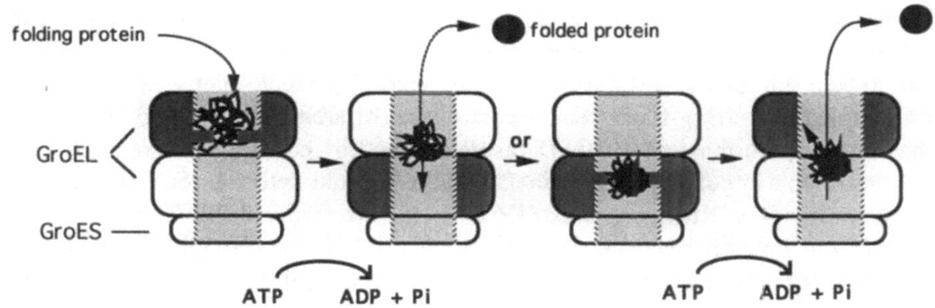

Fig. 3. Hypothetical model for the mechanism of GroEL/GroES-mediated protein folding. See text for description. The two rings of GroEL are drawn in *different shading* to indicate the possibility that they are not functionally equivalent as a consequence of the asymmetry imposed by GroES binding. Alternating of unfolded substrate protein between the two rings has not yet been demonstrated. Not included in the model is the possibility that GroES may transiently dissociate during the folding reaction which takes place within the central cavity of GroEL

GroEL double-ring. This triggers conformational changes in both the GroES adjacent and in the opposite end of the GroEL cylinder, thus apparently lowering the affinity for binding of a second GroES (Langer et al. 1992b).

The 14.5 x 16-nm chaperonin complex has a central cavity about 6 nm in diameter. Bound substrate protein could be detected within this cavity, again by electron microscopy (Langer et al. 1992b; Braig et al. 1993). Based on this finding, we proposed that chaperonin-mediated protein folding may occur while the substrate polypeptide remains in the central space of the GroEL cylinder protected from unproductive intercactions with other folding polypeptides (see also Hendrick and Hartl 1993). The central cavity of GroEL could indeed provide space for proteins of up to 90 kDa (Braig et al. 1993). A substrate protein could make multiple contacts with the GroEL subunits of one seven-member ring, and folding would be achieved by the cooperative, ATP-dependent release of these binding elements. By regulating the K^+-dependent ATPase of GroEL, GroES would coordinate this function (Viitanen et al. 1990; Martin et al. 1991). Figure 3 shows a hypothetical model for the reaction of GroEL/GroES-mediated protein folding. GroES may be transiently released during folding, thus allowing cycles of coordinated ATP-hydrolysis in GroEL which result in substrate release for folding (unpubl. observ.).

6 General Importance of Chaperonin Function

Components functionally related to the chaperonins have recently been discovered in the cytosol of archebacteria and of eukaryotic cells (Phipps et al. 1991; Trent et al. 1992; Lewis et al. 1992; Yaffe et al. 1992; Gao et al. 1992; Frydman et al. 1992). The characterization of the archebacterial chaperonin TF55 had shown that the sequence of TF55 is highly homologous to the sequences of mouse and yeast t-complex polypeptide 1 (Tcp1), a ubiquitous cytosolic protein in eukaryotes (Trent et al. 1992). This suggested that TF55 and Tcp1 belong to a new class of molecular chaperones. We therefore addressed the question whether the Tcp-1 protein of the eukaryotic cytosol has indeed chaperonin-like properties. Tcp-1 was purified from bovine testis and was found to be a subunit of a soluble, hetero-oligomeric high molecular weight complex. This complex had a ring-shape reminiscent of that of archebacterial TF55 and of the chaperonins, and a mass of ~970 kDa as determined by scanning transmission electron microscopy in collaboration with J. Wall. We could detect at least six different subunits in this Tcp-1 ring complex (TRiC) in the size range of 52-65 kDa. Internal peptides from four of these subunits were sequenced. The surprising result was that all of these subunits were related to TCP-1, thus apparently forming a new protein family. Functional characterization showed that TRiC has an ATPase activity comparable to that of GroEL/Hsp60 and is able to bind several unfolded proteins, including firefly luciferase and α and β tubulin. Tubulin was tested because Ursic and Culbertson (1991) had shown that a yeast mutant affecting Tcp-1 was defective in mitotic spindle formation. Tubulin may well be a major substrate of TRiC (Yaffe et al. 1992). TRiC was indeed able to support the ATP-dependent folding of firefly luciferase and the folding/assembly of tubulin that was competent to assemble into microtubules. This first succesful renaturation of unfolded tubulin together with our other results indicates that TRiC indeed has a chaperonin-like function. TRiC and the chaperonins are not functionally identical, however. We did not obtain evidence that TRiC requires the function of a cochaperonin such as GroES. In contrast to TRiC, GroEL did not support the renaturation of luciferase. The unfolded protein bound to GroEL but could not be released even in the presence of Mg-ATP and GroES. These functional differences between TRiC and GroEL/Hsp60 may be related to the hetero-oligomeric composition of TRiC and possibly to the larger size of the ring-complex. A Tcp-1 related protein has also been shown to mediate the folding/assembly of actin (Gao et al. 1992). It is presently unclear whether TRiC participates generally in protein-folding reactions in the cytosol or functions more specifically in the biogenesis of cytoskeletal components.

In summary, the principle of a functional cooperation of different classes of molecular chaperones may be of general importance for in vivo protein folding. The exact sequence of steps, as well as the regulation of these cellular folding pathways, remains to be established. Of particular interest is the molecular mechanism by which double-ring chaperonins mediate protein folding in ATP-dependent reactions. Progress here will depend to a large extent on the availability of structural information for these fascinating proteins.

Acknowledgments. Work in the author's laboratory has been supported by the Deutsche Forschungsgemeinschaft and the National Institutes of Health.

References

Anfinsen CB (1973) Principles that govern the folding of proteins chains. Science 181:223–230

Beckmann RP, Mizzen LA, Welch WJ (1990) Interaction of hsp70 with newly synthesized proteins: implications for protein folding and assembly. Science 248:850–854

Braig K, Simon M, Furuya F, Hainfield JF, Horwich AL (1993) Gold-labeled DHFR binds in the center of GroEL. Proc Natl Acad Sci 90:3978–3982

Buchner J, Schmidt M, Fuchs M, Jaenicke R, Rudolph R, Schmid FX, Kiefhaber T (1991) GroE facilitates refolding of citrate synthase by suppressing aggregation. Biochemistry 30:1586–1591

Cannon S, Wang P, Roy H (1986) Inhibition of ribulose bisphosphate carboxylase assembly by antibody to a binding protein. J Cell Biol 103:1327–1335

Cheng MY, Hartl FU, Martin J, Pollock RA, Kalousek F, Neupert W, Hallberg EM, Hallberg RL, Horwich AL (1989) Mitochondrial heat-shock protein hsp60 is essential for assembly of proteins imported in yeast mitochondria. Nature 337:620–625

Cheng MY, Hartl FU, Horwich AL (1990) Hsp60, the mitochondrial chaperonin, is required for its own assembly. Nature 348:455–458

Christensen H, Pain RH (1991) Molten globule intermediates and protein folding. Eur Biophys J 19:221–229

Creighton TE (1990) Protein folding. Biochem J 270:1–16

Ellis RJ (1987) Proteins as molecular chaperones. Nature 328:378–379

Fischer G, Schmid FX (1990) The mechanism of protein folding. Implications of in vitro refolding models for de novo protein folding and translocation in the cell. Biochemistry 29:2206–2212

Frydman J, Nimmesgern E, Erdjument-Bromage H, Wall JS, Tempst P, Hartl FU (1992) Function in protein folding of TRiC, a cytosolic ring-complex containing TCP1 and structurally related subunits. EMBO J 11:4767–7478

Gao Y, Thomas O, Chow RL, Lee G-H, Cowan NJ (1992) A cytoplasmic chaperonin that catalyzes beta-actin folding. Cell 69:1043–1050

Georgopoulos C, Hendrix RW, Casjens SR, Kaiser AD (1973) Host participation in bacteriophage lambda head assembly. J Mol Biol 76:45–60

Georgopoulos C, Ang D, Liberek K, Zylicz M (1990) Properties of the Escherichia coli heat shock proteins and their role in bacteriophage I growth. In: Morimoto R, Tissieres A, Georgopoulos C (eds) Stress proteins in biology and medicine. Cold Spring Harbor Laboratory, Cold Spring Harbor NY

Gething M-J, Sambrook J (1992) Protein folding in the cell. Nature 355:33–45

Goloubinoff P, Christeller JT, Gatenby A, Lorimer GH (1989b) Reconstitution of active dimeric ribulose bisphosphate carboxylase from an unfolded state depends on two chaperonin protein and Mg-ATP. Nature 342:884–889

Hartl FU, Martin J, Neupert W (1992) Protein folding in the cell: The role of molecular chaperones Hsp70 and Hsp60. Annu Rev Biophys Biomol Struct 21:293–322

Hartman DJ, Hoogenraad NJ, Condron R, Hoj PB (1992) Identification of a mammalian 10-kDa heat shock protein, a mitochondrial chaperonin 10 homologue essential for assisted folding of trimeric ornithine transcarbamoylase in vitro. Proc Natl Acad Sci USA 89:3394–3398

Hemmingsen SM, Ellis RJ (1986) Purification and properties of ribulose bisphosphate carboxylase large subunit binding protein. Plant Physiol 80:269–276

Hemmingsen SM, Woolford C, v.d.Vies SM, Tilly K, Dennis DT, Georgopoulos CP, Hendrix RW, Ellis RJ (1988) Homologous plant and bacterial proteins chaperone oligometric protein assembly. Nature 333:330–334

Hendrick JP, Hartl FU (1993) Molecular chaperone functions of heat-shock proteins. Annu Rev Biochem 62:349–384

Hendrick JP, Langer T, Davis TA, Hartl FU, Wiedmann M (1993) Control of folding and membrane translocation by binding of the chaperone DnaJ to nascent polypeptides (submitted)

Jaenicke R (1991) Protein folding: local structures, domains, subunits, and assemblies. Biochemistry 30:3147–3160

Kang PJ, Ostermann J, Shilling J, Neupert W, Craig EA, Pfanner N (1990) Hsp70 in the mitochondrial matrix is required for translocation and folding of precursor proteins. Nature 348:137–143

Langer T, Lu C, Echols H, Flanagan J, Hayer MK, Hartl FU (1992a) Successive action of molecular chaperones DnaK (Hsp70), DnaJ and GroEL (Hsp60) along the pathway of assisted protein folding. Nature 356:683–689. Lewis et al. (1992)

Langer T, Pfeifer G, Martin J, Baumeister W, Hartl FU (1992b) Chaperonin-mediated protein folding: GroES binds to one end of the GroEL cylinder which accomodates the protein substrate within its central cavity. EMBO J 11:4757–4765

Liberek K, Marszalek J, Ang D, Georgopoulos C, Zylicz M (1991) Escherichia coli DnaJ and GrpE shock proteins jointly stimulate ATPase activity of DnaK. Proc Natl Acad Sci USA 88:2874–2878

Manning-Krieg U, Scherer PE, Schatz G (1991) Sequential action of mitochondrial chaperones in protein import into the matrix. EMBO J 10:3273–3280

Martin J, Langer T, Boteva R, Schramel A, Horwich AL, Hartl FU (1991) Chaperonin-mediated protein folding at the surface of groEL through a "molten globule"-like intermediate. Nature 352:36–42

Martin J, Horwich AL, Hartl FU (1992) Role of chaperonin hsp60 in preventing protein degradation under heat-stress. Science 258:995–998

Mendoza JA, Rogers E, Lorimer GH, Horowitz PM (1991) Chaperonins facilitate the in vitro folding of monomeric mitochondrial rhodanese. J Biol Chem 266:13044–13049

Neupert W, Hartl FU, Craig E, Pfanner N (1990) How do polypeptides cross the mitochondrial membranes? Cell 63:447–450

Ostermann J, Horwich AL, Neupert W, Hartl FU (1989) Protein folding in mitochondria requires complex formation with hsp60 and ATP hydrolysis. Nature 341:125–130

Phipps BM, Hoffmann A, Stetter KO, Baumeister W (1991) A novel ATPase complex selectively accumulated upon heat shock is a major cellular component of thermophilic archaebacteria. EMBO J 10:1711–1722

Scherer PE, Krieg UC, Hwang ST, Vestweber D, Schatz G (1990) A precursor protein partly translocated into yeast mitochondria is bound to a 70 kd mitochondrial stress protein. EMBO J 9:4315–4322

Tandon S, Horowitz PM (1986) Detergent-assisted refolding of gunanidinium chloride-denatured rhodanese. J Biol Chem 261:15615–15681

Trent JD, Nimmesgern E, Wall JS, Hartl FU, Horwich AL (1991) A molecular chaperone from a thermophilic archaebacterium is related to the eukaryotic protein t-complex polypeptide-1. Nature 354:490–493

Ursic D, Culbertson MR (1991) The yeast homolog to mouse Tcp-1 affects microtubule-mediated processes. Mol Cell Biol 11:2629–2640

Viitanen PV, Lubben TH, Reed J, Goloubinoff P, O'Keefe DP, Reed J, Goloubinoff P, O'Keefe DP, Lorimer GH (1990) Chaperonin-facilitated refolding of ribulose bisphosphate carboxylase and ATP hydrolysis by chaperonin 60 (groEL) are K^+dependent. Biochemistry 29:5665–5671

Viitanen PV, Donaldson GK, Lorimer GH, Lubben TH, Gatenby AA (1991) Complex interactions between the chaperonin 60 molecular chaperone and dihydrofolate reductase. Biochemistry 30:9716–9723

Yaffe MB, Farr GW, Miklos D, Horwich AL, Sternlicht ML, Sternlicht H (1992) TCP1 complex is a molecular chaperone in tubulin biogenesis. Nature 358:245–248

MIX
Papier aus verantwortungsvollen Quellen
Paper from responsible sources
FSC® C105338

If you have any concerns about our products,
you can contact us on
ProductSafety@springernature.com

In case Publisher is established outside the EU,
the EU authorized representative is:
Springer Nature Customer Service Center GmbH
Europaplatz 3, 69115 Heidelberg, Germany

Printed by Libri Plureos GmbH
in Hamburg, Germany